Hamid B. Servati

THERMODYNAMIC ANALYSIS OF COMBUSTION ENGINES

Board of Advisors, Engineering

THERMODYNAMIC ANALYSIS OF COMBUSTION ENGINES

ASHLEY S. CAMPBELL
Department of Mechanical Engineering
University of Maine at Orono

JOHN WILEY & SONS
NEW YORK CHICHESTER BRISBANE TORONTO

Library of Congress Cataloging in Publication Data:

Campbell, Ashley S., 1918–
    Thermodynamic analysis of combustion engines.

    Includes index.
    1. Thermodynamics. 2. Heat-engines. 3. Combustion
engineering. I. Title.
TJ265.C25   621.4   78-16181
ISBN 0-471-03751-6

Printed in the United States of America

10 9 8 7 6 5 4 3 2 1

PREFACE

This book is intended as a text for a second course in thermodynamics that treats combustion as the central theme. The objective is to examine the methods of analysis and computational schemes that can be applied to various systems in which motion is produced by combustion.

Most introductory thermodynamics textbooks derive thermal efficiency expressions for those air standard cycles that interest mechanical engineers. This book begins at that point. The first three chapters deal with heat of reaction, adiabatic flame temperature, and isentropic changes of state and provide basic tools. The treatment is approximate, based on the assumption that the composition of the products of combustion can be obtained from the composition of the reactant mixture merely by applying the law of conservation of matter. Succeeding chapters examine gas turbine systems, two- and four-stroke engines, rockets, free piston engines, and boilers and furnaces. The objective is to carry thermodynamic analysis about as far as possible. Here and there performance calculations are presented in tables and plots. Although many more could be added, since the number of assignable operating parameters for combustion engines is large, the tables and plots were selected to illuminate theory and to provide the reader with results against which to compare his or her computations. Details of engine design are discussed only in those instances in which they influence thermodynamic analysis.

The notion of chemical equilibrium is introduced in Chapter 10. The formulation of the governing equations for computing an equilibrium composition is straightforward; however, the subsequent numerical solution is discussed in depth, since the multireaction systems described in others texts have not been treated with what I consider to be adequate detail. Solutions for some simple systems are explained in Chapters 11 and 12.

Chapter 13 reexamines adiabatic flame temperature and isentropic expansion calculations for equilibrium systems. Those circumstances in which chemical equilibrium is likely to have substantial impact on performance computations are outlined in Chapter 14.

Chapter 15 contains a brief treatment of combustion of solid fuels in rockets and powders in guns.

Numbers, and how to derive them, are persistent concerns throughout the book. Engines, at least on paper, come in all manner of size, shape, and design. Operating parameters can be assigned wide ranges of values. These considerations, together with the iterative character of numerical solutions, call for computer programs. None are particularly sophisticated, but all put a high premium on good organization and demonstrate the convenience of function programs and subroutines. Most of the programs are reasonably short. Flow charts have been included only when they explain a chain of calculations with more clarity than a verbal description. Preparing a flow chart is the first order of business in programming and should be left to the reader.

A summary of the SI unit system and a condensed review of ideal gas properties are given in Appendix A. Appendix B describes two popular algorithms for solving equations of the type $F(T) = 0$, which occur repeatedly in combustion problems. Algebraic expressions for computing thermodynamic properties with accuracy sufficient for classroom experience are derived in Appendix C. Tabulated values of these properties, taken from published reports, are listed in Appendix E. Appendix D is a further discussion of chemical equilibrium based on Gibb's free energy, in contrast to the treatment in Chapter 10, which rests on entropy.

I have included only a few problems. Drill questions are out of place, and there is little opportunity for them in any case. I have endeavored to present the material in a manner that will stimulate the reader to formulate questions worth exploring and problems worth solving.

I am deeply obliged to the University of Maine for the sabbatical leave that made it possible for me to draw together a collection of scattered class notes into a coherent form.

ASHLEY S. CAMPBELL

CONTENTS

THERMODYNAMIC ANALYSIS OF COMBUSTION ENGINES

1
THE HEAT OF REACTION

When a fuel burns, the energy associated with the chemical bonds in the fuel and oxygen molecules is released and appears first as a heating effect in the product gases. The temperature ultimately reached in the products of combustion depends on a number of factors: the fuel (or fuels, if more than one is present) and the composition, temperature, and pressure of the reactant mixture. The state of the fuel, whether liquid or gaseous, affects the product temperature. Combustion is a reaction that takes place between gaseous molecules. If the fuel is present as a liquid, it must first be vaporized. The external constraints imposed on the reacting system will affect the product gas temperature. For example, burning at constant pressure and constant volume will lead to different temperatures.

In order to calculate product gas temperatures, we need to know the energy characteristics of fuels, which are variously known as heats of reaction, heating values, heats of combustion, and enthalpies of combustion. In this book, we use the term "heat of reaction." The heat of reaction occupies a position of central importance in combustion thermodynamics and is defined as follows:

When a unit quantity of fuel and chemically correct oxygen burns, the *heats of reaction*, H_{rp} and U_{rp}, represent the energy *added* to bring the product to the initial temperature, 25°C or 298 K, of the reactants, the reaction taking place at constant pressure or constant volume, respectively.

It should be noted that for fuels, H_{rp} and U_{rp} are both negative. (Many reactions have positive heats of reaction. The choice of energy added as the measure of the heat of reaction is merely a convention.) Furthermore, H_{rp} and U_{rp} are properties of the fuel, refer specifically to chemically correct oxygen mixtures reacting at 298 K, and are *measured* quantities.

It will be convenient to define two related energy terms, Q_p and Q_v, as follows:

When a unit quantity of fuel burns, Q_p, and Q_v represent the energy *released* when the products are cooled to the temperature, T, of the reactants, the reaction taking place at constant pressure or constant volume, respectively.

Note that Q_p and Q_v are both positive. They are properties of the *reactant mixture*, refer to any temperature, and are *calculated* from the heats of reaction.

A *chemically correct* (or *stoichiometric*) mixture reacts to produce only CO_2 or H_2O or both. Some examples are given below.

$$CO + .5\ O_2 \rightarrow CO_2$$
$$H_2 + .5\ O_2 \rightarrow H_2O$$
$$CH_4 + 2\ O_2 \rightarrow CO_2 + 2\ H_2O$$
$$C_2H_5OH + 3\ O_2 \rightarrow 2\ CO_2 + 3\ H_2O$$

The constant pressure heat of reaction, H_{rp}, is measured in a continuous flow calorimeter, which is shown in Fig. 1.1, reduced to its bare essentials. Fuel and air in *excess* of the chemically correct requirement enter the reaction chamber at 298 K, and the products are cooled to the inlet temperature by water circulating in the surrounding jacket. For steady flow, conservation of energy requires that

$$H_r(298) + H_w(t_{in}) = H_p(298) + H_w(t_{out}) \qquad \text{1-1}$$

where subscripts r, p, and w refer, respectively, to the reactants, the products, and the cooling water. By definition, H_{rp} is the energy added; hence

$$H_{rp} = M_w c_w(t_{in} - t_{out}) \qquad \text{1-2}$$

where M_w is the mass of water passing through the jacket as unit mass of fuel burns. Clearly, H_{rp} is negative, because t_{in} will be less than t_{out}. Equation 1-2 provides the means for calculating the heat of reaction. It follows that we can write

$$H_{rp} = H_p(298) - H_r(298) \qquad \text{1-3}$$

which is the *thermodynamic definition* or thermodynamic equivalent of H_{rp}.

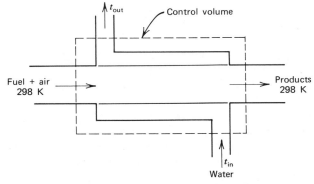

Figure 1.1 Constant flow calorimeter

The feature to be noted is that the reactant mixture contains excess air, not chemically correct oxygen. There are three reasons for performing the measurement in that fashion. First, air is readily available, while oxygen is not. Second, a chemically correct fuel/oxygen mixture would melt the equipment. Third, excess oxygen is used to assure that all the carbon in the fuel will burn to CO_2 and that all the hydrogen will burn to H_2O. But because the reactant and product gas mixtures enter and leave the device at the same temperature, the excess oxygen and the nitrogen contained in the terms on the right-hand side of Eq. 1-3 will cancel out, and it will appear as though the heat of reaction had been measured for a chemically correct fuel/oxygen mixture.

For example, if propane had been burned, then for the two terms on the right-hand side in Eq. 1-3, with x denoting moles of excess oxygen,

$$H_r(298) = [h_{C_3 H_8} + (5+x)h_{O_2} + 3.76(5+x)h_{N_2}]_{298}$$
$$H_p(298) = [3h_{CO_2} + 4h_{H_2 O} + xh_{O_2} + 3.76(5+x)h_{N_2}]_{298}$$

so that for propane,

$$H_{rp}(298) = [3h_{CO_2} + 4h_{H_2 O} - h_{C_3 H_8} - 5h_{O_2}]_{298} \qquad \text{1-4}$$

Equation 1-4 contains the essential combustion feature by which all conceivable combustion configurations for $C_3 H_8$ can be dealt with.

The constant volume heat of reaction, U_{rp}, is measured in a batch calorimeter, illustrated in Fig. 1.2. A measured quantity of fuel, along with excess air, is introduced into a rigid vessel, which is designed to withstand the high pressure that follows after combustion without deforming. (If there is deformation, work will be done on the vessel by the product gases, and a spurious result will be reported.) The vessel is placed in a large tub containing water at room temperature and is left for a period to assure temperature equilibrium and thorough mixing of fuel and oxygen. The fuel/air mixture is then fired, and heat is transferred to the water and vessel walls. The entire system, insulated against heat transfer to the surroundings, comes to temperature

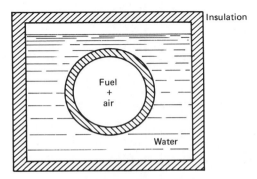

Figure 1.2 Batch calorimeter

equilibrium at some value T', and a correction calculation is made to ascertain the amount of heat transfer necessary to reduce the system again to the initial temperature. Analysis of these events will lead to the companion equation to Eq. 1-3 for U_{rp}, namely

$$U_{rp}(298) = U_p(298) - U_r(298) \qquad \text{1-5}$$

as the thermodynamic definition for U_{rp}. For the case of propane,

$$U_{rp} = [3u_{CO_2} + 4u_{H_2O} - u_{C_3H_8} - 5u_{O_2}]_{298} \qquad \text{1-6}$$

If we know H_{rp}, we can calculate U_{rp}, or vice versa. When Eq. 1-6 is subtracted from Eq. 1-3, we have

$$H_{rp} - U_{rp} = (H_p - U_p) - (H_r - U_r)$$

dropping out the temperature subscript, which is understood always to be 298 K. For gases

$$h - u = RT$$

where R is the universal gas constant, since the enthalpy h and internal energy u are in units per mole. For liquids

$$h - u \approx 0$$

The difference between the constant pressure and constant volume heats of reaction may then be written as

$$H_{rp} - U_{rp} = (N_p - N_r)RT$$

and with $R = 8.314$ kJ/kmol

$$H_{rp} - U_{rp} = 2480 \, (N_p - N_r) \qquad \text{1-7}$$

where N_p and N_r denote, respectively, the mole numbers in the product and reactant mixture of the *gaseous* species for a chemically correct fuel/oxygen mixture. To continue with the propane example, from Table E.4, Appendix E

$$H_{rp} = -2\ 032\ 800 \text{ kJ/kmol propane}$$

for gaseous propane and gaseous H_2O in the products. Then

$$U_{rp} = -2\ 032\ 800 - (7 - 6) \times 2480$$
$$= -2\ 035\ 300 \text{ kJ/kmol propane}$$

or a change of less than 1%.

From Table E.4, the two H_{rp} values are for liquid and gaseous fuel, with gaseous H_2O in both cases. The difference is the heat of vaporization of propane at 298 K, that is,

$$h_{fg} = 2\ 032\ 800 - 2\ 016\ 900$$
$$= 15\ 900 \text{ kJ/kmol propane}$$

The effect of condensing the water vapor in the products on H_{rp} and U_{rp} values may be substantial. If we denote

H_{rp} = constant pressure heat of reaction, gaseous H_2O

H'_{rp} = constant pressure heat of reaction, liquid H_2O

then

$$H_{rp} - H'_{rp} = N_{H_2O} \times h_{fg}(H_2O) \qquad \text{1-8}$$

For propane, with $h_{fg}(H_2O) = 43\ 956$ kJ/kmol,

$$H_{rp} - H'_{rp} = 4 \times 43\ 956 = 175\ 824 \text{ kJ/kmol propane}$$

and

$$H'_{rp} = -2\ 208\ 600 \text{ kJ/kmol propane}$$

a change of more than 8%. Consequently, when securing heat of reaction values from published tables, care must be exercised to note the condition of the water in the products. Frequently the term "lower" heating value is attached to H_{rp} or U_{rp} with gaseous H_2O, and the term "higher" heating value is attached to the case of liquid H_2O.

When a reaction represents the formation of a compound from its elements, the heat of reaction becomes the *heat of formation*. When a compound can be formed from its elements in more than one way, there is opportunity for confusion. For example,

$$H_2 + .5 \, O_2 \rightarrow H_2O, \quad \Delta h_f^0 = -241\,800 \text{ kJ/kmol } H_2O$$

$$H_2 + 0 \quad\;\; \rightarrow H_2O, \quad \Delta h_f = -489\,300 \text{ kJ/kmol } H_2O$$

$$2H + .5 \, O_2 \rightarrow H_2O, \quad \Delta h_f = -677\,700 \text{ kJ/kmol } H_2O$$

$$2H + 0 \quad\;\; \rightarrow H_2O, \quad \Delta h_f = -925\,000 \text{ kJ/kmol } H_2O$$

when the reactions take place at 298 K, and the H_2O formed is gaseous. The first reaction represents the formation of H_2O from its elements in their standard state, that is, the natural state for hydrogen and oxygen at room temperature and pressure (in this case, gases). The heat of formation from elements in the standard state is identified by the superscript ^{0} and is referred to as the *standard heat of formation*, or the *standard enthalpy of formation*. Table E-4 lists a few values for powders which may be of interest in subsequent discussions.

Frequently a heat of formation is included in the reaction equation; for example,

$$CO + .5 \, O_2 \rightarrow CO_2 + 282\,800 \text{ kJ/kmol } CO_2 \text{ at 298 K} \qquad\qquad 1\text{-}9$$

The line is to be read as follows: one kilomole of CO and one-half kilomole of O_2 will produce one kilomole of CO_2 and 282 800 kilojoules, when the reaction proceeds isothermally at 298 K. The reaction is *exothermic*, and the heat of formation is negative; energy must be transferred from rather than to the reaction if it is to proceed isothermally.

Table E.4 lists H_{rp} values for the paraffin hydrocarbon series, the straight-chain molecules, for which *n*-octane is an example

$$
\begin{array}{c}
\text{H \ H \ H \ H \ H \ H \ H \ H} \\
\text{H–C–C–C–C–C–C–C–C–H} \\
\text{H \ H \ H \ H \ H \ H \ H \ H}
\end{array}
$$

When the H_{rp} values listed in the table are divided by the molecular weight, the units are converted to kilojoules per kilogram, with numerical results that differ only slightly from methane to *n*-decane. Constant heat of reaction per unit mass is a characteristic of the hydrocarbon series.

SUMMARY

The heats of reaction, H_{rp} and U_{rp}, represent the energy that must be added to the products of combustion of a unit quantity of fuel in order to bring those products to the initial temperature of the reactants. H_{rp} denotes the energy added for constant pressure combustion, and U_{rp} denotes the energy added for constant volume combustion. The thermodynamic definitions are

$$H_{rp} = H_p(298) - H_r(298)$$

$$U_{rp} = U_p(298) - U_r(298)$$

where p and r denote products and reactants. The reactants are always understood to be a chemically correct fuel/oxygen mixture. The products contain only CO_2 or H_2O or both.

In the next chapter, we examine the manner in which H_{rp} and U_{rp} are employed to calculate adiabatic combustion temperatures.

PROBLEMS

1.1 For propane, C_3H_8, calculate the heat of reaction in kilojoules per kilomole (kJ/kmol) and in calories per kilogram (cal/kg) for the following eight conditions:

Constraint	Fuel	H_2O in Products
Constant volume	Liquid	Liquid
Constant volume	Liquid	Gaseous
Constant volume	Gaseous	Gaseous
Constant volume	Gaseous	Liquid
Constant pressure	Liquid	Liquid
Constant pressure	Liquid	Gaseous
Constant pressure	Gaseous	Gaseous
Constant pressure	Gaseous	Liquid

1.2 For the hydrocarbons and the alcohols, more energy is released when combustion takes place at constant volume than when it takes place at constant pressure.

For gases, and mixtures of gases, more energy is required to heat at constant pressure than to heat at constant volume.

Are these two statements in conflict, and if so, how can they be resolved?

2
THE ADIABATIC FLAME TEMPERATURE

2.1 INTRODUCTION

Calculation of the adiabatic flame temperature occurs in virtually every problem that involves combustion. The techniques for its calculation are discussed in this chapter.

The adiabatic flame temperature depends on a number of factors, the major ones being (1) the fuel or fuels, (2) the chemical composition of the reactant mixture, (3) the temperature of the reactant mixture, (4) the constraints on the system, and (5) the pressure of the reactant mixture. Intuitively, we expect different fuels to produce different flame temperatures, the ratio of hydrogen/carbon atoms being the major parameter. Likewise, flame temperature is expected to vary as reactant mixture composition is varied. Mixture with an overabundance of fuel or air, for example, will produce low flame temperatures. Fuel/oxygen mixtures will produce higher temperatures than fuel/air mixtures with an equivalent oxygen content; the nitrogen present in air is essentially inert, does not contribute to the combustion reaction, but merely absorbs energy. Raising the reactant mixture temperature should increase flame temperature. Reactants burned at constant volume do not transfer work to the surroundings, as do reactants burned at constant pressure. These two constraint systems will therefore produce different flame temperatures. Reactant pressure can affect flame temperatures; this is explained in detail in Chapter 10.

Accurate flame temperature computations involve a complicated system of equations. The discussion in this chapter will be confined to a simpler set of equations for two reasons. First, the accurate calculation tends to obscure the physics of what is taking place. Second, the approximate flame temperature computation frequently produces predictions that are "good enough," although we cannot say how good until approximate and accurate temperatures can be compared. These comparisons are discussed in Chapter 13.

The temperature calculation is carried out by trial and error, if done by hand, or by a systematic search, if done by computer. The emphasis here will be on organization of the solution for computer programming.

2.2 COMPLETE COMBUSTION IN C/H/N/O SYSTEMS

Under the scheme known as *complete combustion*, the composition of the product gas mixture can be obtained merely by inspecting the reactant mixture. For a reactant mixture that contains C, H, N, and O atoms, the rules are simply as follows:

1. All nitrogen appears as N_2; that is, nitrogen is inert.
2. All hydrogen appears as H_2O.
3. All carbon is oxidized to CO. If any oxygen remains, part of the CO is oxidized to CO_2.
4. If there is sufficient oxygen to oxidize all the carbon to CO_2, the excess appears as O_2.

(A similar scheme is available for reactant mixtures containing H, N, and O atoms. This is discussed in Section 2.8.)

The product gas mixture mole numbers can now be set down for the general case. Let *MC*, *MH* and *MO* denote, respectively, the number of moles of carbon, hydrogen, and oxygen *atoms* in a mole of fuel. Let *YCC* denote the chemically correct amount of oxygen per mole of fuel. Then

$$YCC = MC + MH/4 - MO/2 \qquad\qquad 2\text{-}1$$

Let *YMIN* denote the minimum allowable oxygen content in the reactants per mole of fuel, so that

$$YMIN = (MC - MO)/2 + MH/4 = YCC - MC/2 \qquad\qquad 2\text{-}2$$

For a reactant mixture containing

$$1 \text{ mole of fuel} + Y\,O_2 + 3.76Y\,N_2$$

the product mole numbers are

$YMIN \leqslant Y \leqslant YCC$	$Y \geqslant YCC$	
$N(1) = 2(YCC - Y)$	$N(1) = 0$	
$N(2) = 2(Y - YMIN)$	$N(2) = MC$	
$N(3) = MH/2$	$N(3) = MH/2$	2-3
$N(4) = 3.76\ Y$	$N(4) = 3.76\ Y$	
$N(5) = 0$	$N(5) = Y - YCC$	

where we adopt the FORTRAN subscripting system, so that correspondence between mole numbers and gases is

$$1 = CO, \qquad 2 = CO_2, \qquad 3 = H_2O, \qquad 4 = N_2, \qquad 5 = O_2 \qquad\qquad 2\text{-}4$$

With this general framework, we now examine the two important flame temperature calculations, namely, adiabatic combustion at constant pressure and at constant volume. We shall develop the system of equations for a particular fuel, $C_{10}H_{22}$. The equations are then less cumbersome, and we can pass easily to the general formulation of the solution for any fuel.

2.3 CONSTANT PRESSURE ADIABATIC COMBUSTION
The problem is shown in Fig. 2.1. The reactants

$$C_{10}H_{22} + Y O_2 + 3.76Y N_2$$

enter an insulated combustion chamber at 298 K, and we wish to know the temperature T of the products. The energy equation is

$$H_r(298) = H_p(T) \qquad\qquad 2\text{-}5$$

The definition of the heat of reaction for $C_{10}H_{22}$ is

$$H_{rp} = [10h_{CO_2} + 11h_{H_2O} - h_{C_{10}H_{22}} - 15.5h_{O_2}]_{298} \qquad\qquad 2\text{-}6$$

Consider first the case, $Y \geqslant YCC$; then

$$H_r(298) = [h_{C_{10}H_{22}} + Yh_{O_2} + 3.76Yh_{N_2}]_{298} \qquad\qquad 2\text{-}7$$

$$\cdot\ H_p(T) = [10h_{CO_2} + 11h_{H_2O} + 3.76Yh_{N_2} + (Y - 15.5)h_{O_2}]_T \qquad\qquad 2\text{-}8$$

These four equations are now manipulated as follows: substitute Eqs. 2-7 and 2-8 into 2-5; then solve Eq. 2-6 for the term $h_{C_{10}H_{22}}$ and eliminate that term from Eq. 2-5. The result will be

$$-H_{rp} = 10\,\Delta h_{CO_2} + 11\,\Delta h_{H_2O} + 3.76Y\,\Delta h_{N_2} + (Y - 15.5)\,\Delta h_{O_2} \qquad\qquad 2\text{-}9$$

Figure 2.1 Constant pressure adiabatic combustion

where

$$\Delta h_i = h_i(T) - h_i(298) \qquad\qquad 2\text{-}10$$

Note that the left-hand side of Eq. 2-9 is a measured quantity, and the right-hand side contains pairs of enthalpy differences.

Now consider the other case, $YMIN \leqslant Y \leqslant YCC$. Equations 2-5, 2-6, and 2-7 are unchanged. In place of Eq. 2-8, we now have

$$H_p(T) = [2(15.5 - Y)h_{CO} + 2(Y - 10.5)h_{CO_2} + 11h_{H_2O} + 3.76Yh_{N_2}]_T \quad 2\text{-}11$$

Equation 2-11 replaces Eq. 2-8, and we carry out the same manipulations, which lead to

$$-H_{rp} = 11\,\Delta h_{H_2O} + 3.76Y\,\Delta h_{N_2} + 2(15.5 - Y)h_{CO}(T) + 2(Y - 10.5)h_{CO_2}(T)$$
$$+ (15.5 - Y)h_{O_2}(298) - 10h_{CO_2}(298) \qquad\qquad 2\text{-}12$$

We need to bring Eq. 2-12 into a form that contains pairs of enthalpy differences only on the right-hand side. We can do this by writing the following terms:

$$2(15.5 - Y)h_{CO}(298)$$

$$2(Y - 10.5)h_{CO_2}(298)$$

on the right-hand side, once positive and once negative. We can then regroup and obtain

$$-H_{rp} - 2(15.5 - Y)(h_{CO} + \tfrac{1}{2}h_{O_2} - h_{CO_2})_{298}$$
$$= 2(15.5 - Y)\,\Delta h_{CO} + 2(Y - 10.5)\,\Delta h_{CO_2} + 11\,\Delta h_{H_2O} + 3.76Y\,\Delta h_{N_2} \quad 2\text{-}13$$

The essential step now is to recognize the group

$$(h_{CO} + \tfrac{1}{2}h_{O_2} - h_{CO_2})_{298}$$

as the heat of reaction for the dissociation of CO_2. That reaction has a positive heat of reaction amounting to 282 800 kJ/kmol CO_2, a value that can be found in chemical handbooks or in Eq. 1-9.

Now compare Eqs. 2-9 and 2-13, the results so far for the two cases for Y. Since $N(1) = 0$ for $Y \geqslant YCC$, we could add the term

$$N(1)\,\Delta h_{CO}$$

on the right-hand side of Eq. 2-9, and the equation remains valid. Similarly, when $YMIN \leqslant Y \leqslant YCC, N(5) = 0$, so the term

$$N(5) \Delta h_{O_2}$$

can be added to the right-hand side of Eq. 2-13 without altering its validity. When we have done this, Eqs. 2-9 and 2-13 take the form

$$Q_p = N(1) \Delta h_{CO} + N(2) \Delta h_{CO_2} + N(3) \Delta h_{H_2O} + N(4) \Delta h_{N_2} + N(5) \Delta h_{O_2}$$

or simply

$$Q_p = \Sigma N(I) \Delta h_I, \qquad I = 1,2,3,4,5 \qquad \text{2-14}$$

and

$$Q_p = -H_{rp} - N(1) \, 282 \, 800$$

where $\Delta h_I = h_I(T) - h_I(298)$.

Equation 2-14 is the final result. Note that the summation extends over all five product gases. Equation 2-14, together with the formulations of the mole numbers in Eqs. 2-3, provides the solution for T, the adiabatic flame temperature.

Figure 2.2 illustrates the physical interpretation of Eq. 2-14. The right-hand side represents the energy required to heat the products of combustion from 298 K to T. As shown in the figure, the reactants enter the reactor at 298 K, react, and are cooled to 298 K, which releases energy Q_p, which is now routed to the heater, so that the requirement of adiabatic combustion will be preserved. The control volume around the heater will lead to Eq. 2-14. As defined in Chapter 1 and repeated here,

Q_p represents the energy *released* when a unit quantity of fuel is burned at constant pressure and the products of combustion are cooled to the initial temperature of the reactants. It is positive, because H_{rp} is negative, and it is a property of the reactant mixture, since $N(1)$, the moles of CO in the products, depends on the value of Y, which in turn characterizes the reactant mixture.

Some idea of the variation of T with Y can be obtained by writing Eq. 2-14 in the form

$$Q_p = N_p \int_{298}^{T} c_p \, dT \qquad \text{2-15}$$

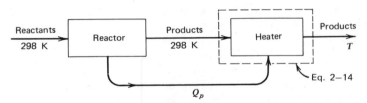

Figure 2.2 Physical description of Eq. 2-14

where N_p is the number of moles of products, and c_p is the heat capacity of the product mixture. From Eqs. 2-3 for the product mole numbers, for $C_{10}H_{22}$,

$$N_p = 21 + 3.76Y, \qquad Y \leqslant YCC$$

$$Np = 5.5 + 4.76Y, \qquad Y \geqslant YCC$$

If the fuel enters the reaction as a gas, and the H_2O in the products is a gas, $H_{rp} = -6\,312\,300$ kJ and

$$Q_p = 6\,312\,300 - N(1)\,282\,800 \text{ kJ}$$

We then have the following

Y	N_p	$N(1)$	Q_p	Q_p/N_p
11.0	62.4	9.0	3 767 100	60 400
12.5	68.0	6.0	4 615 500	67 900
14.0	73.6	3.0	5 463 900	74 200
15.5	79.3	0	6 312 300	79 600
17.0	86.4	0	6 312 300	73 100
18.5	93.6	0	6 312 300	67 400
20.0	100.1	0	6 312 300	63 100

The energy available for heating the products, per mole of product mixture, falls off on either side of the chemically correct mixture. Since c_p in Eq. 2-15 varies slowly with Y, as compared to the variation of $N(1)$ with Y, the curve of T versus Y can be expected to follow closely the variation of Q_p/N_p.

Equation 2-14 is a general equation, and we see that what characterizes a combustion system, when treated from the simple *complete combustion* scheme, are the values of H_{rp}, MC, MH, MO, and Y.

2.4 CONSTANT VOLUME ADIABATIC COMBUSTION

The problem is shown in Fig. 2.3. Reactants at temperature 298 K and pressure P_r are ignited, and both the flame temperature T and the final pressure P_p are to be calculated. The analysis follows the same sort of manipulations described for the case of constant pressure combustion. The size of the vessel is of no concern, since pressure and temperature are intensive quantities, independent of the size of the system. We can conveniently choose to work with a reactant mixture containing a mole of fuel. The energy equation is now

$$U_r(298) = U_p(T)$$

For $C_{10}H_{22}$, if selected as the fuel, the heat of reaction is defined as

$$U_{rp} = [10u_{CO_2} + 11u_{H_2O} - u_{C_{10}H_{22}} - 15.5u_{O_2}]_{298} \qquad \text{2-16}$$

For the reactant mixture

$$H_r(298) = [u_{C_{10}H_{22}} + Yu_{O_2} + 3.76Yu_{N_2}]_{298} \qquad \text{2-17}$$

Expressions are now written for $U_p(T)$ for the two Y regimes and are manipulated as before. The group

$$(u_{CO} + \tfrac{1}{2}u_{O_2} - u_{CO_2})_{298}$$

will appear. From Eq. 1-7, we find U_{rp} for the dissociation of CO_2,

$$U_{rp} = H_{rp} - 2480 (N_p - N_r)$$
$$= 282\ 800 - 2480\ (1.5 - 1)$$
$$= 281\ 400 \text{ kJ/kmol } CO_2$$

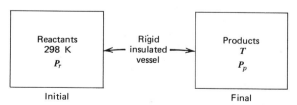

Figure 2.3 Constant volume adiabatic combustion

and the counterpart to Eq. 2-14 will read

$$Q_v = \Sigma N(I) \, \Delta u_I, \qquad I = 1,2,3,4,5$$

with   2-18

$$Q_v = -U_{rp} - N(1) \, 281 \; 400$$

where $\Delta u_I = u_I(T) - u_I(298)$.

Equation 2-18 is the general expression for any constant volume adiabatic combustion problem. The right-hand side represents the energy necessary to heat the products from 298 K to T. The physical interpretation of Eq. 2-18 is illustrated in Fig. 2.4. The reactants are burned, and the products are immediately cooled to 298 K, with the release of energy Q_v, which is promptly returned to the products, since the entire process must be adiabatic.

Q_v represents the energy *released* when a unit quantity of fuel is burned at constant volume, and the products of combustion are cooled to the initial temperature of the reactants. It is positive, because U_{rp} is negative, and is a property of the reactant mixture, since $N(1)$, the number of moles of CO in the products, depends on the value of Y, which in turn characterizes the reactant mixture.

The final pressure, P_p, is obtained from

$$P_p = P_r \frac{N_p T}{N_r \, 298}$$

where N_r denotes the moles of gaseous reactants.

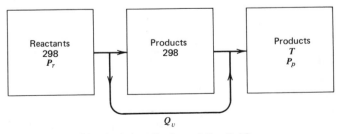

Figure 2.4 Physical description of Eq. 2-18

2.5 NUMERICAL SOLUTION FOR THE FLAME TEMPERATURE

Consider first the case of constant pressure combustion, Eq. 2-14. Since the enthalpy increases with temperature, and the slope also increases with temperature, the Newton-Raphson iterative scheme provides a rapidly converging, systematic search for the flame temperature. Rewrite Eq. 2-14 in the form

$$\Sigma \left[N(I)h_I(T) - N(I)h_I(298) \right] - Q_p = 0 \qquad \text{2-19}$$

We need the derivative of that function with respect to T, which is simply

$$\Sigma N(I)c_{p,I}(T) \qquad \text{2-20}$$

the heat capacity of the product mixture at T. (See Appendix B for a brief discussion of the Newton-Raphson form of solution.)

Appendix C describes the formulation of algebraic expressions for the enthalpy and heat capacity of gases. Expressions for the enthalpy, for example, take the form

$$h(T) = AL + BL\,T + CL\,\ln(T) \qquad 400 < T < 1600$$
$$h(T) = AH + BH\,T + CH\,\ln(T) \qquad 1600 < T < 6000 \qquad \text{2-21}$$

and the heat capacity is then

$$c_p(T) = BL + CL/T \qquad 400 < T < 1600$$
$$c_p(T) = BH + CH/T \qquad 1600 < T < 6000 \qquad \text{2-22}$$

Values for the coefficients $AL, BL, \ldots, CH$ for each gas are listed in Table C.1 of Appendix C. (There are two more coefficients, DL and DH, which occur in the entropy expression.) A data card is prepared for each gas, containing the eight coefficients and the enthalpy of the gas at 298 K (call it HO), obtained from Table E.2, Appendix E. Denote the enthalpy of the product gas mixture by the function $HP(T)$. The following function program will prove convenient:

```
FUNCTION HP(T)
REAL N
COMMON N(5),AL(5),BL(5),CL(5),AH(5),BH(5),CH(5)
HP=0
IF(T.GT.1600.) GO TO 2
DO 1 I=1,5
1  HP=HP+N(I)*(AL(I)+BL(I)*T+CL(I)*ALOG(T))
RETURN
2  DO 3 I=1,5
3  HP =HP+N(I)*(AH(I)+BH(I)*T+CH(I)*ALOG(T))
RETURN
END
```

2-23

Since the expression for enthalpy, Eq. 2-21, does not extend to 298 K, the enthalpy of the product mixture at 298 K, call it HP 298, can be evaluated by

```
   HP298=0
   DO 4 I=1,5
 4 HP298=HP298+N(I)*HO(I)
```
2-24

The Newton-Raphson recursion expression can be given the form

```
   TNEW=T-(HP(T)-HP298 -QP)/CP(T)
```
2-25

A function program for $CP(T)$ can be written, which follows the form for $HP(T)$. The iterative scheme then reduces to the following set of statements, which begin with an arbitrary guess value for T:

```
   T-2000.
 5 TNEW=T-(HP(T)-HP298-QP)/CP(T)
   IF(ABS(T-TNEW).LT.5.) GO TO 6
   T=TNEW
   GO TO 5
 6 CONTINUE
```
2-26

and calculates the flame temperature to within ± 5 degrees.

The calculation for the constant volume adiabatic flame temperature follows the same pattern. In place of Eq. 2-25, we have, from Eq. 2-18,

```
   TNEW=T-(UP(T)-UP298-QV)/CV(T)
```
2-27

Function programs could be written for $UP(T)$ and $CV(T)$, or use could be made of the relations

```
   UP = HP(T) - 8.314*NP*T (kJ)
```
2-28

```
   CV = CP(T) - 8.314*NP (kJ/K)
```

where NP is the sum of the product mole numbers.

2.6 THE ADIABATIC FLAME TEMPERATURE: GENERAL CASE

In the general case the reactants are at T_r, not 298 K, as in Sections 2.3 and 2.4. When the analysis of the general case is carried through, we discover that Q_p and Q_v values are required at T_r. Since Q_p and Q_v both involve heats of reaction, we inquire into the variation of H_{rp} and U_{rp} with temperature. The difference can be ignored, except for large changes in temperature. For H_{rp}, for example,

$$H_{rp}(T_r) - H_{rp}(298) \approx (N_p c_p - N_r c_p)_{cc} (T_r - 298) \qquad \text{2-29}$$

where the subscript cc identifies the quantities within the bracket as those that apply to the chemically correct reaction. Usually, T_r differs from 298 by only a few hundred degrees; furthermore, the differences between the mole number and heat capacity products is seldom large. As a consequence, it is legitimate to ignore the temperature variations in H_{rp} and U_{rp}.

Then, for the general case of constant pressure adiabatic combustion, with reactants at T_r, the flame temperature is obtained from

$$Q_p = \Sigma N(I)[h_I(T) - h_I(T_r)]$$

where   2-30

$$Q_p = -H_{rp} - N_{CO} \, 282 \, 800$$

and for constant volume adiabatic combustion

$$Q_v = \Sigma N(I)[u_I(T) - u_I(Tr)]$$

where   2-31

$$Q_v = -U_{rp} - N_{CO} \, 281 \, 400$$

In both equations, the summation can be extended over *all* the product gases.

For the general case, the recursion expressions in the Newton-Raphson iteration appear now, for constant pressure combustion, as

TNEW=T−(HP(T)−HP(TR)−QP)/CP(T)   2-32

and, for constant volume combustion, as

TNEW=T−(UP(T)−UP(TR)−QV)/CV(T)   2-33

The last two expressions indicate the convenience of function programs for $HP(T)$, $UP(T)$, $CP(T)$, and $CV(T)$.

2.7 THE ADIABATIC FLAME TEMPERATURE: AN ALTERNATE METHOD

Thermodynamics defines enthalpy and internal energy in terms of differentials, dh and du. In order to tabulate h or u values against temperature, a datum temperature must be selected and an arbitrary base value assigned to either h or u. These choices may be made independently for each gas in the tabulation.

It is advantageous to assign the base values in accordance with the measured heats of formation. In constructing tabulations of, say, h values, the values for the elements

$$C, H_2, O_2, \text{ and } N_2$$

may be chosen arbitrarily, since no element may be formed from other elements, and assigned at 298.16 K. Then the h values of the gases

$$CO, CO_2, H, H_2O, OH, O, \text{ and } N$$

are assigned, again at 298.16 K, from the known values of the heats of formation for those seven gases. With a set of tabulated enthalpy values computed in this fashion (Table E.2, is such a set) the solution of the adiabatic flame temperature proceeds in a manner different from what we have seen so far.

As an example, let the combustion take place between gaseous $C_{10}H_{22}$ and air, both at 298 K. As before, the energy equation for constant pressure combustion relates the reactant and product mixture enthalpies

$$H_r(298) = H_p(T)$$

The heat of reaction equation is

$$H_{rp} = (10h_{CO_2} + 11h_{H_2O} - h_{C_{10}H_{22}} - 15.5h_{O_2})_{298} \qquad \text{2-34}$$

Now solve this expression for the fuel term, using values from Table E.2 at 298 K,

$$h_{CO_2} = 9\ 364 \text{ kJ/kmol}$$
$$h_{H_2O} = 57\ 316$$
$$h_{O_2} = 17\ 200$$
$$h_{N_2} = 15\ 780$$

and from Table E.4,

$$H_{rp} = -6\ 312\ 300 \text{ kJ/kmol}$$

evaluate the enthalpy of the fuel

$$h_{C_{10}H_{22}}(298) = 6\ 769\ 800\ \text{kJ/kmol} \qquad \text{2-35}$$

Then

$$H_r(298) = (h_{C_{10}H_{22}} + Yh_{O_2} + 3.76Yh_{N_2})_{298} \qquad \text{2-36}$$
$$= 6\ 769\ 800 + 76\ 530Y$$

and the energy equation is

$$6\ 769\ 800 + 76\ 530Y = \Sigma N(I)h_I(T), \qquad T_r = 298\ \text{K} \qquad \text{2-37}$$

which may be solved by a Newton-Raphson iteration for T.

For the case of constant volume combustion of gaseous $C_{10}H_{22}$ and air, the energy equation is

$$U_r(298) = U_p(T)$$

when the reactants are at 298 K. From Eq. 1-7

$$U_{rp} = H_{rp} - 2480\,(N_p - N_r)$$
$$= -6\ 312\ 800 - 2480\,(21 - 16.5)$$
$$= -6\ 323\ 500\ \text{kJ/kmol}$$

At 298 K

$$u_{CO_2} = 6\ 886\ \text{kJ/kmol}$$
$$u_{H_2O} = 54\ 838$$
$$u_{O_2} = 14\ 722$$
$$u_{N_2} = 13\ 302$$

From the definition of U_{rp},

$$U_{rp} = (10u_{CO_2} + 11u_{H_2O} - u_{C_{10}H_{22}} - 15.5u_{O_2})_{298} \qquad \text{2-38}$$

evaluate the internal energy of $C_{10}H_{22}$ at 298 K,

$$u_{C_{10}H_{22}}(298) = 6\ 767\ 400\ \text{kJ/kmol}$$

The internal energy of the reactants is

$$U_r(298) = (u_{C_{10}H_{22}} + Yu_{O_2} + 3.76Yu_{N_2})$$
$$= 6\ 767\ 400 + 64\ 740Y$$

and the energy equation, transformed into a working equation for T, reads

$$6\ 767\ 400 + 64\ 740Y = \Sigma N(I)u_I(T), \qquad T_r = 298\ \text{K} \qquad \text{2-39}$$

The point here, of course, is that Eqs. 2-37 and 2-39, developed by this alternative approach, are obtained without the sort of manipulations that were necessary in Sections 2.3 and 2.4. (The advantage of the alternative method will become evident when the full solution, with the product gas mixture in chemical equilibrium, is discussed in Chapter 13.)

Section 2.6 dealt with the general case for reactants at some temperature T_r different from 298 K. We can do the same for the alternative method. For constant pressure combustion,

$$H_r(T_r) = H_p(T)$$

For $H_r(T_r)$ we can write

$$H_r(T_r) = H_r(298) + \int_{298}^{T_r} C_p\ dT$$

where C_p is the heat capacity of the reactant mixture,

$$C_p = c_p(C_{10}H_{22}) + Yc_p(O_2) + 3.76Yc_p(N_2) \qquad \text{2-40}$$

The working form for the equation which leads to the adiabatic flame temperature for constant pressure combustion is then

$$6\ 769\ 800 + 76\ 530Y + \int_{298}^{T_r} C_p\ dT = \Sigma N(I)h_I(T) \qquad \text{2-41}$$

When T_r is greater than 298 K, the integral will be positive; increasing the reactant temperature increases the flame temperature, as is to be expected.

The constant volume counterpart to Eq. 2-41 is

$$6\,767\,400 + 64\,740Y + \int_{298}^{T_r} C_v \, dT = \Sigma N(I) u_I(T) \qquad \text{2-42}$$

where C_v is the constant volume heat capacity of the reactant mixture.

Table E.1 lists values for the coefficients a and b in a heat capacity expression

$$c_p = a + bT \qquad 298 < T < 900$$

for various gases that may occur in the reactant mixture of a combustion problem. The expression is intended to be simple, for use in the integrals of Eq. 2-41 and 2-42.

As a closing comment, in the case of constant pressure combustion, the fuel and air may enter the combustion chamber at different temperatures, T_{fuel} and T_{air}. For this condition, the integral in Eq. 2-41 should be split in two parts, separating the fuel and air terms and carrying each integration to its appropriate upper limit.

Although Eqs. 2-30 and 2-41 apply to the same problem (constant pressure combustion) and yet have quite different appearances, both will lead to identical numbers for the flame temperature. The same comment may be made about Eqs. 2-31 and 2.42. Whichever equations are selected for the solution, the mole numbers for the product gases are given by Eqs. 2-3, which are functions of Y, moles of oxygen per mole of fuel.

Constant pressure combustion, with fuel and oxidizer at different temperatures, may not have the sort of complications that Eq. 2-41 appears to suggest. For example, rockets can be fired with a combination of liquid $C_{10}H_{22}$ (which is a reasonable approximation of kerosene) and liquid oxygen. Suppose, as illustrated in Fig. 2.5, the fuel is pumped in at 298 K and the oxygen at 90 K. As in the previous examples, let Y denote the moles of oxygen burned per mole of fuel. Then the energy equation is

$$H_r = H_p(T)$$

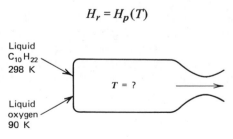

Figure 2.5 Rocket motor

and we know that

$$H_r = h_{C_{10}H_{22}}(\text{liq}, 298) + Yh_{O_2}(\text{liq}, 90) \qquad\qquad \text{2-43}$$

From the thermodynamic definition of H_{rp} for gaseous $C_{10}H_{22}$ we found (Eq. 2-35)

$$h_{C_{10}H_{22}}(\text{gas}, 298) = 6\ 769\ 800 \text{ kJ/kmol}$$

The heat of vaporization for $C_{10}H_{22}$ is found from the H_{rp} values, Table E.4,

$$h_{fg}(298) = 6\ 312\ 300 - 6\ 261\ 300 = 51\ 000 \text{ kJ/kmol}$$

so that

$$
\begin{aligned}
h_{C_{10}H_{22}}(\text{liq}, 298) &= 6\ 769\ 800 - 51\ 000 \\
&= 6\ 718\ 800 \text{ kJ/kmol}
\end{aligned}
$$

The oxygen term in Eq. 2-43 comes directly from Table E.2,

$$h_{O_2}(\text{liq}, 90) = 4\ 315 \text{ kJ/kmol}$$

The energy equation takes the final form

$$6\ 718\ 800 + 4\ 315\,Y = \Sigma N(I)h_I(T) \qquad\qquad \text{2-44}$$

Note that in the product mole number expressions, Eq. 2-3, $N(4) = 0$, since there is no nitrogen in this combustion system.

2.8 COMPLETE COMBUSTION IN H/N/O SYSTEMS

The product mole numbers for H/N/O systems are obtained by inspection of the reactants, assuming that all nitrogen will be N_2, that excess oxygen beyond the chemically correct requirement will appear as O_2, and that unburned hydrogen will appear as H_2. With these assumptions, the product mole numbers for the reactant mixture

$$H_2 + Y\,O_2 + 3.76Y\,N_2 \qquad\qquad \text{2-45}$$

will be

$0 \leqslant Y \leqslant .5$	$Y \geqslant .5$
$N(1) = 1 - 2Y$	$N(1) = 0$
$N(2) = 2Y$	$N(2) = 1$
$N(3) = 3.76Y$	$N(3) = 3.76Y$
$N(4) = 0$	$N(4) = Y - .5$

2-46

where the correspondence between subscript I and the gases is

$$1 = H_2, \qquad 2 = H_2O, \qquad 3 = N_2, \qquad 4 = O_2$$

Two cases are examined. The results are given, and the proof is left to the reader as an exercise.

For constant volume adiabatic combustion of hydrogen and air (as in an internal combustion engine, e.g.) the energy equation will reduce to

$$Q_v = \Sigma N(I)[u_I(T) - u_I(T_r)]$$

and

$$Q_v = 241\ 900\ [1 - N(1)]$$

2-47

Some indication of the variation between T and Y can be obtained by writing Eq. 2-47 in the form

$$Q_v = N_p \int_{T_r}^{T} c_v\ dT$$

in which N_p is the total moles of products, and c_v is the heat capacity of the products. We can then calculate Q_v/N_p for various Y values as follows:

Y	N_p	$N(1)$	Q_v	Q_v/N_p
.2	1.75	.6	96 760	55 300
.3	2.13	.4	145 100	68 100
.4	2.50	.2	193 500	77 400
.5	2.88	0	241 900	84 000
.6	3.36	0	241 900	72 000
.7	3.83	0	241 900	63 000

which suggests that the flame temperature will reach its maximum value at, or at least close to, the chemically correct reactant mixture composition.

For constant pressure adiabatic combustion of liquid hydrogen and liquid oxygen (as in a rocket motor), the combustion chamber temperature, T, is obtained from

$$282\ 612 + 4\ 315Y = \Sigma N(I)h_I(T) \qquad\qquad 2\text{-}48$$

with $N(3) = 0$ in Eq. 2-46.

2.9 SUMMARY

Calculation of the adiabatic flame temperature, as discussed in this chapter, proceeds from the basic assumption that we can assign a composition to the product gas mixture that depends only on the composition of the reactant mixture. Examples are given for C/H/N/O systems in Eqs. 2-3, and for H/N/O systems in Eqs. 2-46. The flame temperatures thus calculated are *higher* than the actual values found in practice, since they do not take into account the energy absorbed by large molecules as they dissociate to smaller molecules and that absorbed by small molecules as they dissociate to atoms. This subject is discussed in Chapter 10.

Two methods are outlined with which the energy equation may be reduced to a working equation for the computation of the flame temperature. The method discussed in Sections 2.3, 2.4, and 2.6 leads to equations of the type shown in Eqs. 2-14, 2-18 and Eqs. 2-30, 2-31. The right-hand side in each of those expressions contains paired enthalpy or internal energy *differences*. Consequently, those equations, and any others derived by manipulating the sets of equations that define the flame temperature problem, may be solved with *any* set of enthalpy and internal energy tables, since the datum values for h and u, however selected, will cancel out.

An alternate method for securing a working equation for the flame temperature computation is outlined in Section 2.7 and leads to equations of the type shown in Eqs. 2-37, 2-39, 2-41, 2-42, 2-44, and 2-48. In those equations, enthalpy and internal energy terms are unpaired, so that *special* tables (Table E.2 is one example) are required, since datum values for h and u will not cancel.

In all of these equations, the summations on the right-hand side can extend over *all* the product species. (Of course, some mole numbers will be zero.) This arrangement simplifies the organization of a computer program to do the flame temperature computation. The Newton-Raphson iteration, discussed in Section 2.5, provides a systematic search for the flame temperature.

PROBLEMS

Problems 2.1 through 2.8 should be carried out by hand. In the process of doing so, the reader will find that hand calculations are a tedious business. The purpose of these eight problems is to reveal this fact, as well as to acquaint the reader with the order of magnitude of adiabatic flame temperatures. When engine performance is examined later in the book, the desirability of computer programs that compute flame temperatures will become evident.

2.1 Compute the flame temperature when a chemically correct mixture of gaseous propane C_3H_8 and air, initially at 300 K and 1 atm, burn at constant pressure. Do the problem two ways, using the methods in Section 2.3 and 2.7.

2.2 Repeat Problem 2.1 for liquid propane.

2.3 Repeat Problem 2.1 for constant volume combustion, and compute the final pressure.

2.4 Compute the flame temperature for constant pressure combustion for a methane CH_4/air mixture, initially at 300 K and 1 atm, that contains 20% more air than is required for chemically correct conditions.

2.5 Repeat Problem 2.4, but change the initial temperature of the mixture to 800 K.

2.6 Compute the flame temperature for constant pressure combustion of liquid methyl alcohol CH_3OH mixed with 92% of chemically correct air, with reactants at 300 K and 1 atm.

2.7 Various fuels, principally alcohols, that can be derived by distilling organic matter, wood for example, are viewed as possible substitutes for petroleum-based fuels. Methanol/gasoline mixtures are one possibility. Suppose methyl alcohol CH_3OH is combined with octane C_8H_{18}, the former supplying 69% of the mass, and is injected as a liquid, at 300 K, into an airstream that is at 800 K and carrying 90% of the chemically correct air. What will be the resulting flame temperature, if combustion proceeds at constant pressure?

2.8 Does the presence of water vapor in the air supporting combustion play a significant role? To answer the question, repeat Problem 2.1 with saturated air.

2.9 For the gases CO, CO_2, H_2, H_2O, N_2, and O_2, punch a data card for each gas that contains the eight coefficients, $AL, BL, \ldots, DH$, found in Table C.1, Appendix C, and the enthalpy at 298.16 K, found in Table E.1.)

Write a program that reads each data card in turn and computes c_p and h over the range 500–6000 K, at 500-degree increments, using Eqs. 2-21 and 2-22. Compare the printout with tabulated values in Tables E.1 and E.2. (You will discover that the low temperature c_p values show poor agreement. This results from using such a simple expression as Eq. 2-22. But there is no cause for alarm. For the most part, we are interested in c_p values at high temperature. Low temperature values, when needed, can be evaluated from the equations in Table E.1.)

After reading in each data card, and before making the calculations, print out the information on the data card. It is a good habit to read out any numbers that have been read in; it provides a check on the card itself and also on the read format.

It is desirable to identify the appropriate gas for each card. This can be accomplished by punching the chemical symbol in the first two or three columns. The following format can be employed to read (and write) a card:

A4,F8.0,F6.2,F9.1,F8.2,F8.0,F6.2,F9.1,F8.2,F8.0,F4.0

A card for carbon monoxide will appear as follows:

CO 299180. 37.85 4571.9 −31.1 309070. 39.29 −6201.9 −42.77 283807. 28.

The eight numbers that follow "CO" are the low and high temperature coefficients. The next number is h at 298 K, and the last number is the molecular weight.

2.10   This exercise provides experience with one of the essential building blocks in the construction of programs for engine performance analysis.

Construct a computer program that will compute the flame temperature for constant pressure combustion of gaseous $C_{10}H_{22}$/air mixtures, with reactants at 298 K. The program may be written to follow the method of Section 2.3 or 2.7. Arrange the program to perform the computation for a range of fuel/air ratios.

2.11   With a small change in the program of Problem 2.10, the flame temperature for constant volume combustion of liquid $C_{10}H_{22}$/air mixtures, with reactants at 298 K, can be carried out for a range of fuel/air ratios.

2.12   Return to the program of Problem 2.10 and remove the nitrogen. Note the enormous change in flame temperature. This explains why heats of reaction are measured using fuel/air mixtures rather than fuel/oxygen mixtures.

2.13   Arrange the program of Problem 2.10 to compute the flame temperature for constant pressure combustion of liquid hydrogen/liquid oxygen mixtures. For this combination of reactants, Section 2.7 provides the easy method. This problem can be used for part of the solution to Problem 3.2 in the next chapter.

2.14   For all the gases listed in Table E.1, plot c_p against T, for the full range of 300 to 6000 K. Plot c_p as the ordinate, with the scale extending to zero at the T-axis. When plotted in this fashion, you can quickly visualize the percentage error in the enthalpy difference between any two temperatures when the temperature variation in c_p is ignored.

3
ISENTROPIC CHANGES OF STATE

Every type of combustion engine involves an expansion process. Many involve compression as well. The working fluid is a mixture of gases. The problem is always to calculate the temperature of the gases at the end of the process of compression or expansion. In piston/cylinder engines, the final temperature enables us to calculate the work done by or on the gases. The same applies to compressors and gas turbines. The exit temperature in a rocket motor nozzle enables us to calculate the rate of momentum change of the exhaust gases and the reaction force that they exert on the motor.

Ideally, the compressions and expansions take place rapidly and without transfer of heat, that is, reversibly and adiabatically. Hence, compression and expansion calculations are based on the assumption of isentropic behavior. Where they are known not to be isentropic, as in compressors and turbines, the isentropic temperature computation is still required, because efficiencies are based on isentropic performance.

The relationships between pressure, volume and temperature for an ideal gas, or mixture of ideal gases, at the same entropy are

$$\frac{T_2}{T_1} = \left(\frac{P_2}{P_1}\right)^{(k-1)/k} = \left(\frac{V_1}{V_2}\right)^{k-1}, \qquad k = \frac{c_p}{c_v} \qquad \text{3-1}$$

These expressions are derived (see Appendix A for a summary of ideal gas properties) by an integration that assumes k is constant. In fact, of course, c_p and c_v increase with temperature, while k decreases. Since c_p and c_v differ by a constant amount, the change in k with temperature is not as strong, however. Nevertheless, the application of Eqs. 3-1 may lead to error in the final temperature. The error is larger for expansions from a high temperature than for compressions from a low temperature. Hence, we shall concentrate our attention on the expansion process.

Let P_1 and T_1 denote the initial state of a mixture of gases. We wish to determine the temperature, T_2, when the mixture expands isentropically to P_2. (We shall look at this problem first and then examine the case of expansion between known volumes, V_1 and V_2.) There are three ways to proceed.

First, we can apply Eq. 3-1, find T_2, and let it go at that.

Second, we can modify the error of the first method by finding an average k value. For example, with k_1 found at T_1, we find a T_2 value and the corresponding value k_2. Now form an average between k_1 and k_2, which should lead to a better approximation for T_2. This routine can be repeated until successive T_2 values differ from one another by whatever amount we choose to call small. But how do we form the average between k_1, which remains fixed, and k_2, which varies with each new value for T_2? Should we use an arithmetic average

$$\frac{k_1 + k_2}{2} \qquad\qquad 3\text{-}2$$

or a geometric average

$$\sqrt{k_1 k_2}$$

or what? There is no apparent logic to guide us.

Third, we can solve for T_2 correctly by finding a value that equates the initial and final entropies of the gas mixture. The total entropy $S(P,T)$ of a mixture of ideal gases is given by

$$S(P,T) = \Sigma n_i \phi_i(T) - NR \ln(P) \qquad\qquad 3\text{-}3$$

where n_i denotes mole numbers; N, total mole number; R, the universal constant; and ϕ_i, a function of temperature.

$$\phi_i = \int_0^T \frac{c_{p_i}}{T} \, dT + \phi_i(0) \qquad\qquad 3\text{-}4$$

where $\phi_i(0)$ is an integration constant. For any gas, $\phi_i(T)$ is the entropy per mole at temperature T and 1 atm. Table E.3 tabulates values of ϕ for the gases that commonly occur in combustion products. Appendix C lists in Table C.1 the coefficients that can be used in the following expressions:

$$\phi = BL \ln(T) - CL/T + DL \qquad 400 < T < 1600$$
$$\qquad\qquad 3\text{-}5$$
$$\phi = BH \ln(T) - CH/T + DH \qquad 1600 < T < 6000$$

Then T_2 is found by solving the equation

$$\Delta S = S(P_2, T_2) - S(P_1, T_1) = 0 \qquad\qquad 3\text{-}6$$

which can be expanded to

$$\Delta S = \Sigma \, n_i \left[\phi_i(T_2) - \phi_i(T_1) - NR \, \ln\left(\frac{P_2}{P_1}\right) \right] = 0 \qquad\qquad 3\text{-}7$$

It is convenient to provide the computer program with a function program that generates values for $S(P, T)$ for given values of P, T and the mole numbers $N(I)$. For a system of five gases, the following is an example. The total mole number is NT.

```
      FUNCTION S(P,T)
      REAL N,NT
      COMMON NT,N(5),BL(5),CL(5),DL(5),BH(5),CH(5),DH(5)
      S=0
      IF(T.GT.1600.) GO TO 2                                    3-8
      DO 1 I=1,5
   1  S=S+N(I)*(BL(I)*ALOG(T)−CL(I)/T+DH(I))
      GO TO 4
   2  DO 3 I=1,5
   3  S=S+N(I)*(BH(I)*ALOG(T)−CH(I)/T+DH(I))
   4  S=S−8.314*NT*ALOG(P)
      RETURN
      END
```

Equation 3-6 may be solved for T_2 by a half-interval search. For $T_2 = T_1$ we know ΔS is positive, since P_2 is less than P_1. The procedure is then to reduce T_2 by some fixed amount until ΔS changes sign. At that point the solution is bracketed, and the half-interval search can go into operation.

The solution of Eq. 3-7 by the Newton-Raphson iteration is more rapid. To write the recursion expression, we need the derivative of ΔS with respect to T_2,

$$\frac{d}{dT_2} (S_2 - S_1) = \frac{C_p(T_2)}{T_2}, \qquad C_p = \Sigma \, n_i c_{p\,i} \qquad\qquad 3\text{-}9$$

It is convenient to incorporate a function program for $CP(T)$ in the program. The recursion equation would then appear as the following statement:

```
      T2NEW=T2*(1.−(S(P2,T2)−S(P1,T1))/CP(T2))              3-10
```

Some care must be exercised because the program will abort when $T2NEW$ turns up negative as a result of the logarithm function in Eq. 3-5. Figure 3.1 illustrates the

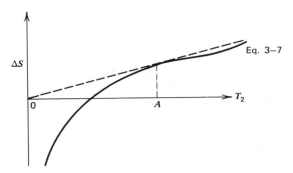

Figure 3.1 Newton-Raphson iteration for Eq. 3-7. The guess value for T_2 must be less than A.

problem. The slope of ΔS plotted against T_2 will decrease as shown because T_2 increases at a faster rate than C_p. The limiting value for initial guess at T_2 is at A in the figure. The program will abort if the iteration begins with a larger value. Since we do not know the limiting value at the outset, we can avoid trouble by choosing any T_2 to begin the iterations and then arrange to decrease T_2 by some fixed amount until a positive *T2NEW* is obtained from the recursion statement. Thereafter in the iterations, *T2 NEW* will remain positive.

For the case of isentropic expansion to some specified volume V_2, the same basic program of iterations can be employed, P_2 being fixed by

$$P_2 = \frac{P_1 V_1}{T_1 V_2} T_2 \qquad\qquad 3\text{-}11$$

SUMMARY

The material in Chapters 1, 2, and 3 provide the requisite concepts and techniques for dealing with a large variety of combustion problems. This chapter raised a question that is as yet unanswered: how should isentropic expansions be calculated? There is no single answer to such a broad question. It is suggested that Problem 3.2 at the end of this chapter be completed, as it will provide hints about the answer.

PROBLEMS

3.1   Return to the program in Problem 2.9 and add in the computation of ϕ, the temperature portion of the entropy function, given in Eqs. 3-5. For each gas, make the computation over the range of 500 to 6000 K, and compare the results with the tabulated values of ϕ in Table E.3, Appendix E.

3.2   Liquid hydrogen and liquid oxygen are continuously pumped into a combustion chamber, burn adiabatically at a constant pressure of 30 atm, and then expand through a shock-free nozzle to 1 atm. Compute the exit velocity for a range of reactant mixtures.

Computation of the exit velocity involves the temperature of the products at the exit section of the nozzle. As outlined in the text, this can be accomplished in three ways:

(*a*) By a single pass, using Eq. 3-1.
(*b*) By using an average k value, based on the k values at the combustion chamber temperature, and the temperature reached in method a.
(*c*) By using the isentropic relationship in Eq. 3-7. Make the velocity calculation for each method.

Combustion is employed for either of two purposes: to provide a high temperature environment or to produce motion. The following three problems involve motion resulting directly from combustion.

3.3 A piston with mass M rests on stops a distance z above the closed end of a cylinder, length L and cross-sectional area A, open at the top and closed at the bottom (see Fig. 3.2). The cylinder is held in a vertical position. The space below the piston is filled with a gaseous propane C_3H_8/air mixture, at 100 kPa and 25°C. The ambient pressure is 100 kPa, and the device is at sea level. The propane/air mixture is ignited and burns rapidly. The piston is blown out of the cylinder and rises to a height H above the base of the cylinder.

The problem contains two quantities of interest: the height H and the efficiency of the device, which can be defined as

$$\text{Efficiency (\%)} = -\frac{Mg(H-z)}{mU_{rp}} \times 100$$

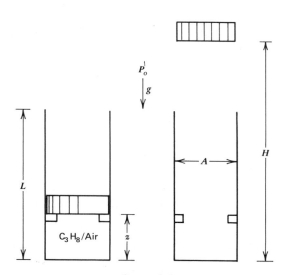

Figure 3.2

where m is the quantity of propane burned, and U_{rp} is the constant volume heat of reaction. (Note that the device is charged with U_{rp}, not Q_v, the heat released by combustion.)

In the initial position, before combustion, the device contains two independent parameters: z, the piston position; and the fuel/air ratio. We can expect H and the efficiency to be dependent on these two parameters. For

$$M = 5 \text{ kg}, \qquad L = 3 \text{ m}, \qquad A = 80 \text{ cm}^2$$

for what values of z and fuel/air ratio will

(a) H reach its maximum value?
(b) Efficiency reach its maximum value?

To simplify the problem, assume the products of combustion expand according to $Pv^k = $ constant, which is supported by the results of Problem 3.2. Also, assume that combustion occurs instantaneously and ignore friction and leakage at the piston.

3.4   Return to Problem 3.3 and include a friction force, $FR = 75$ N, between the piston and cylinder wall. Find the conditions for maximum H and maximum thermal efficiency.

3.5   Usually, the starting or breakaway friction force is larger than the sliding friction force. Return to Problem 3.4 and let the starting friction force be 125 N and the sliding friction force be 75 N. Find the conditions for maximum H and maximum thermal efficiency.

3.6   A piston with mass M and area A is positioned distance L from the left-hand end of a horizontal tube, closed at both ends, and distance LL from the right-hand end (see Fig. 3.3). A chemically correct mixture of C_3H_8/air is introduced to the left of the piston, with air to the right. The propane/air mixture is ignited and burns very rapidly. There is no leakage between piston and cylinder and no friction. For

$$M = 10 \text{ kg}, \qquad L = 0.3 \text{ m}, \qquad LL = 3.0 \text{ m}, \qquad A = 0.05 \text{ m}^2$$

find the time required for the piston to make one to-and-fro oscillation.

3.7   The assumptions regarding friction and leakage in Problem 3.6 are inconsistent. If there is no friction, there must be leakage of gas from one side of the piston to the other; if there is no leakage, there must be friction.

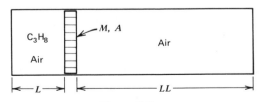

Figure 3.3

Leakage is rather difficult to deal with, because the direction of flow changes with piston position. On the other hand, friction is not difficult to include in the problem. When it is included, we can observe how the motion decays with time.

Repeat Problem 3.6 with a friction force $FR = 400$ N. Follow the motion until the piston stops. Assume that product gases and air each expand and contract according to $Pv^k = $ constant. The specific heat ratios for the two are not, of course, the same. The positions in the tube when the piston stops and reverses motion must be found by some trial-and-error scheme. You should give some thought to the allowable error in this computation.

3.8   A fuel/air mixture at P_o, T_o is ignited and burns adiabatically at constant volume. A valve in the containing vessel is opened, and the product gases expand rapidly to P_{atm}. What factors in this situation influence the mass fraction of products that will remain in the vessel?

4
GAS TURBINE CYCLES

4.1 INTRODUCTION

In this chapter we examine the performance of several gas turbine configurations from the point of view of thermodynamics, which means studying the effects of various parameters on thermal efficiency and on net work output.

As the performance of compressors, particularly axial flow compressors of the type found in aircraft propulsion units, improves, and as new advances in materials and metallurgy increase the safe operating temperature of turbine blades and buckets, new applications for gas turbines seem to abound. The simplicity of power production with gas turbines, as compared to steam turbines, accounts in large part for this growing interest.

4.2 THE AIR STANDARD BRAYTON CYCLE

Two examples of the Brayton cycle, the open and closed cycles, are shown in Fig. 4.1. The closed cycle comprises compression, heating, expansion, and cooling. More work is produced during expansion than is required for compression, so that net work can be delivered to the surroundings. In the open cycle, air is compressed, the temperature is increased by combustion, and products are discharged to the surroundings, along with net work. The open cycle is not a true cycle.

Certain essential features of both forms of work production can be illustrated on a temperature-entropy diagram. It is helpful to have one drawn to scale. From Appendix A, the entropy of a gas, in terms of temperature and pressure, is

$$ds = c_p \frac{dT}{T} - R \frac{dP}{P} \qquad \text{4-1}$$

Assuming c_p constant, which simplifies the algebra but does not obscure the information we are after,

$$s = c_p \ln(T) - R \ln(P) + \text{constant}$$

Set $s(300 \text{ K}, 20 \text{ atm}) = 0, R = 8.314 \text{ kJ/kmol K}$. Take $c_p = 30 \text{ kJ/kmol K}$. We can then obtain the following values for s:

		s — entropy kJ/kmol K		
	P (atm)			
T	1	5	10	20
300	25.1	11.8	6.0	0
500	40.5	27.1	21.4	15.4
700	50.6	37.3	31.5	25.5
900	58.2	44.9	39.1	33.1
1100	64.2	50.9	45.1	39.1
1300	69.2	55.9	50.1	44.1
1500	73.5	60.2	54.5	48.4

The four constant pressure lines are shown in Fig. 4.2. The lines have positive slope, which increases with T because

$$\left(\frac{\partial T}{\partial s} \right)_P = \frac{T}{c_p} \qquad\qquad 4\text{-}2$$

Furthermore, the slope is a function of T only. (This holds true even if c_p is not constant, because c_p is not dependent on pressure until high pressures are reached.) Consequently, if we have a template shaped to any constant pressure line, other lines can be generated simply by moving the template horizontally. The distance between constant pressure lines is constant when measured parallel to the s-axis, but it increases with T when measured parallel to the T-axis. This is the key property of gases which makes the Brayton cycle feasible as a work producer. Note the uniform spacing between the 5 and 10 and the 10 and 20 atm lines. Why?

Figure 4.2 shows three cycles: 1-2'-3'-4', 1-2"-3"-, and so forth. Each cycle is anchored to point 1, the compressor inlet (300 K, 1 atm), and the turbine inlet temperature is 1300 K in each cycle. That figure represents the top operating limit for turbines today, a limit fixed by the strength of the turbine blades. Comparing the lengths of the expansion lines with their corresponding compression lines, it is evident that each cycle will deliver work to the surroundings.

To examine the performance of these ideal cycles, let

$$w_c = \text{work done on the gas during compression}$$
$$= h_2 - h_1 = c_p(T_2 - T_1)$$
$$w_t = \text{work done by the gas during expansion}$$
$$= h_3 - h_4 = c_p(T_3 - T_4) \qquad\qquad 4\text{-}3$$
$$Q_a = \text{heat added to the gas}$$
$$= h_3 - h_2 = c_p(T_3 - T_2)$$

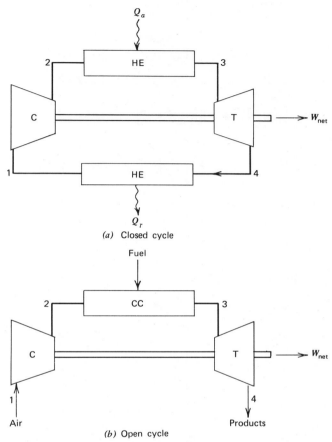

Figure 4.1 Two examples of the simple Brayton cycle. C = compressor, T = turbine, HE = heat exchanger, CC = combustion chamber, Q_a = energy added, Q_r = energy rejected.

Each process is a steady flow event. If the velocities are low, the temperatures are stream temperatures; if high, they represent stagnation temperatures. The ideal cycles operate without pressure drops; hence,

$$P_2 = P_3 \qquad \text{and} \qquad P_1 = P_4$$

It is convenient to introduce the following piece of notation:

$$PR = \left(\frac{P_2}{P_1}\right)^{(k-1)/k} = \left(\frac{P_3}{P_4}\right)^{(k-1)/k} \qquad \text{4-4}$$

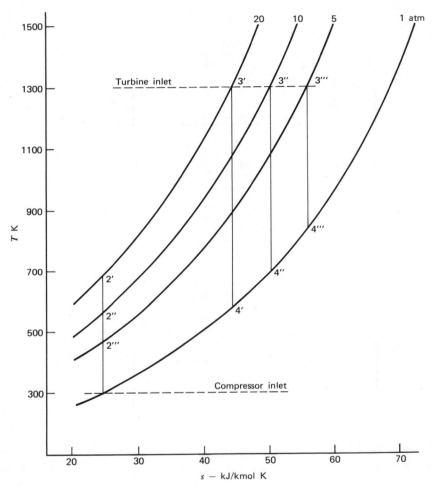

Figure 4.2 Temperature-entropy diagram for an ideal gas, c_p = 30 kJ/kmol K.

Then

$$T_2 = PRT_1 \qquad \text{and} \qquad T_4 = T_3/PR \qquad\qquad 4\text{-}5$$

so that

$$w_c = c_p T_1(PR - 1) \qquad \text{and} \qquad w_t = c_p T_3(1 - 1/PR) \qquad\qquad 4\text{-}6$$

For the thermal efficiency, η_{th}, we have

$$\eta_{th} = \frac{\text{net work done}}{\text{heat added}}$$

$$= \frac{(T_3 - T_4) - (T_2 - T_1)}{(T_3 - T_2)} \qquad \text{4-7}$$

and since

$$PR = \frac{T_2}{T_1} = \frac{T_3}{T_4}$$

we have

$$\frac{T_4}{T_1} = \frac{T_3}{T_2}$$

and the thermal efficiency reduces to

$$\eta_{th} = 1 - \frac{T_1}{T_2} = 1 - \frac{1}{(P_2/P_1)^{(k-1)/k}} = 1 - \frac{1}{PR} \qquad \text{4-8}$$

Thermal efficiency increases as the pressure ratio across the compressor (or turbine) increases. Another feature of interest and, as it turns out, concern is the work ratio, W_r.

$$W_r = \frac{w_c}{w_t} = \frac{T_1}{T_3} \left(\frac{P_2}{P_1}\right)^{(k-1)/k} \qquad \text{4-9}$$

which, for fixed inlet temperatures to the compressor and turbine, also increases with pressure ratio. What this means is that for some specified amount of net work output, $w_t - w_c$, we require high pressure ratio for high thermal efficiency, which will mean large compressor and large turbine.

Turbines and compressors cannot be built to operate at 100% efficiency. The efficiencies are defined as follows:

$$\eta_c = \text{compressor efficiency}$$

$$= \frac{\text{isentropic work input between } (P_1, T_1) \text{ and } P_2}{\text{actual work input between } (P_1, T_1) \text{ and } P_2} \qquad \text{4-10}$$

$$\eta_t = \text{turbine efficiency}$$

$$= \frac{\text{actual work output between } (P_3, T_3) \text{ and } P_4}{\text{isentropic work output between } (P_3, T_3) \text{ and } P_4} \qquad \text{4-11}$$

Compressor manufacturers, for example, announce the efficiency of their products by measuring the work input between some specified inlet pressure and temperature and some specified outlet pressure. The isentropic work input between the same inlet and outlet conditions can be calculated and the efficiency computed. The efficiency of a compressor may be a function of all three — inlet pressure, inlet temperature, and outlet pressure. Turbine manufacturers do the same measuring and computing.

Since compressors and turbines are steady flow devices in the Brayton cycle,

$$\eta_c = \frac{h_{2s} - h_1}{h_2 - h_1} \qquad \text{and} \qquad \eta_t = \frac{h_3 - h_4}{h_3 - h_{4s}} \qquad\qquad 4\text{-}12$$

The s subscript denotes the isentropic state associated with the inlet condition to the machine. The actual compressor outlet temperature, T_2, and the actual turbine outlet temperature, T_4, are now functions of the efficiencies

$$T_2 = T_1 \left[1 + \frac{PR - 1}{\eta_c} \right] \qquad\qquad 4\text{-}13$$

$$T_4 = T_3 \left[1 - \eta_t \frac{PR - 1}{PR} \right] \qquad\qquad 4\text{-}14$$

To demonstrate the rapid decrease in thermal efficiency and the corresponding increase in work ratio, let $\eta_c = \eta_t$ and with

$$T_1 = 300 \text{ K}, \qquad T_3 = 1300 \text{ K}, \qquad c_p = 30 \text{ kJ/kmol K}$$

$$k = \frac{30}{30 - 8.314} = 1.38, \qquad P_2 = 10P_1, \qquad PR = 10^{.277} = 1.89$$

calculate T_2 and T_4 from Eqs. 4-13 and 4-14. The results are as follows:

$\eta_c = \eta_t$	T_2	T_4	η_{th}	W_r	$T_4 - T_2$
1.0	568	687	.47	.44	119
.9	598	749	.36	.54	151
.8	635	810	.23	.68	175
.7	683	871	.07	.89	188

Thermal efficiency is computed from Eq. 4-7. The result given in Eq. 4-8 applies only for 100% efficient machines.

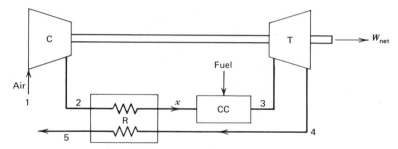

Figure 4.3 Open Brayton cycle, with regenerator. C = compressor, T = turbine, CC = combustion chamber, R = regenerator.

As long as the gas temperature leaving the turbine exceeds the gas temperature entering the heater, or the combustion chamber, there is an opportunity to improve thermal efficiency by transferring energy from the turbine discharge to the compressor discharge. But this opportunity is not always present. Figure 4.2 shows that at 5 atm, T_4''' is greater than T_2'''; but at 20 atm, T_2' exceeds T_4'. For every combination of T_1, T_3, η_c, and η_t, there is an upper limit on P_2/P_1 beyond which this desirable heat transfer cannot take place.

Figure 4.3 shows the open cycle with a heat exchanger added. It is called a *regenerator* and operates in countercurrent fashion. Under the best conditions, air leaving the regenerator would have been heated to T_4. When we set $T_2 = T_4$, the thermal efficiency for an *ideal* cycle with *complete regeneration* becomes

$$\eta_{th} = 1 - \left(\frac{T_1}{T_3}\right)\left(\frac{P_2}{P_1}\right)^{(k-1)/k} \qquad\qquad 4\text{-}15$$

which bears some resemblance to the equation for the work ratio, W_r. Is this coincidence? See Eq. 4-9.

By equating Eqs. 4-8 and 4-15, we find the pressure ratio for which the regenerated and unregenerated cycles give identical thermal efficiencies,

$$\frac{P_2}{P_1} = \left(\frac{T_3}{T_1}\right)^{k/2(k-1)} \qquad\qquad 4\text{-}16$$

Note that Eq. 4-16, like Eqs. 4-8 and 4-15, applies only to cycles operating with *ideal* (100% efficient) compressors and turbines. Because of the manner in which it was derived, Eq. 4-16 states the pressure ratio beyond which the cycle cannot incorporate a regenerator; that is, for larger P_2/P_1, we will always find $T_2 > T_4$.

So much for the ideal cycles. In the next section we examine the performance of stationary turbines that incorporate combustion and regeneration.

4.3 THE BRAYTON CYCLE AS A GAS TURBINE POWER PLANT

The notation follows Fig. 4.3, which incorporates a regenerator. The compressor inlet conditions, P_1 and T_1, are fixed by local conditions. We are free to choose the following parameters:

P_2 = compressor outlet pressure

η_c = compressor efficiency

η_t = turbine efficiency

T_3 = turbine inlet temperature

The fuel

ϵ = regenerator effectiveness

$$= \frac{T_x - T_2}{T_4 - T_2} \qquad\qquad 4\text{-}17$$

The pressure ratio across the turbine is now

$$\frac{P_3}{P_4} = \frac{P_2 - \Delta P}{P_1 + \Delta P} \qquad\qquad 4\text{-}18$$

where ΔP is the pressure drop that accompanies the air and product gas flows through the regenerator. (A pressure drop across the combustion chamber could also be included and that will further reduce P_3.) Using the notation introduced in Eq. 4-4,

$$PR_a = \left(\frac{P_2}{P_1}\right)^{(k_a-1)/k_a}, \qquad PR_t = \left(\frac{P_3}{P_4}\right)^{(k_p-1)/k_p} \qquad\qquad 4\text{-}19$$

where subscripts a and p denote, respectively, air and products. The equations that describe the operation of the power plant and lead to numbers for the thermal efficiency and net work output are

$$T_2 = T_1\left[1 + \frac{PR_a - 1}{\eta_c}\right]$$

$$w_c = c_{pa}(T_2 - T_1) \qquad\qquad 4\text{-}20$$

The next three form a group,

$$T_x = T_2 + \epsilon(T_4 - T_2) \qquad\qquad 4\text{-}21$$

$$H_r(T_x) = H_p(T_3) \qquad \text{4-22}$$

$$T_4 = T_3 \left[1 - \eta_t \left(1 - \frac{1}{PR_t} \right) \right] \qquad \text{4-23}$$

Continuing,

$$w_t = c_{pp}(T_3 - T_4)(1 + F/A) \qquad \text{4-24}$$

$$w_{net} = w_t - w_c$$

$$\eta_{th} = \frac{w_{net}}{-H_{rp}} \frac{A}{F} \qquad \text{4-25}$$

To find T_x in Eq. 4-21, we need T_4; and to find T_4 from Eq. 4-23, we need ΔP and k_p. The value for ΔP should be associated in some way with regenerator effectiveness ϵ. On the other hand, k_p depends on the composition of the product gases, which in turn depends on the solution of Eq. 4-22, which describes constant pressure adiabatic combustion.

A simple iterative routine will get us out of this circle. Assume for the moment that we have some sort of relationship between ΔP and ϵ, so that by assigning a number to the latter we know the value of the former. Then

1. Assume a value for k_p (1.30 is a good value)
2. Calculate PR_t from Eqs. 4-18 and 4-19
3. Calculate T_4 from Eq. 4-23
4. Calculate T_x from Eq. 4-21
5. Calculate the product gas composition from Eq. 4-22. This will lead to a k_p value.
6. Compare the assumed and calculated k_p values; if they compare favorably, we are done; otherwise the calculated k_p is inserted in step 1, and the iterations continue. (The convergence is very rapid.)

The factor F/A in Eqs. 4-24 and 4-25 represents fuel/air ratio on a *mass* basis. (More mass flows through the turbine than through the compressor.) Consequently, the proper units for c_{pa} and c_{pp} are kilojoules per kilogram Kelvin (kJ/kg K). Then w_{net} will be expressed in kilojoules per kilogram of air, and the heat of reaction H_{rp} in Eq. 4-25 must have units of kilojoules per kilogram of fuel.

Turning to the energy equation for adiabatic combustion, Eq. 4-22, when we ignore the temperature difference (if any) between air and fuel entering the combustion chamber, the equation takes the form of Eq. 2-30.

$$-H_{rp} = \Sigma N(I)[h_I(T_3) - h_I(T_x)] \qquad \text{4-26}$$

When the fuel is $C_{10}H_{22}$, for example, the reactants are

$$C_{10}H_{22} + Y O_2 + 3.76Y N_2$$

and the products will be

$$10CO_2 + 11H_2O + (Y - 15.5)O_2 + 3.76Y N_2$$

so that the energy equation may be solved at once for Y,

$$Y = \frac{-H_{rp} - 10\Delta h_{CO_2} - 11\Delta h_{H_2O} + 15.5\Delta h_{O_2}}{\Delta h_{O_2} + 3.76\Delta h_{N_2}} \qquad \text{4-27}$$

where

$$\Delta h_i = h_i(T_3) - h_i(T_x).$$

When expressions for the h_i terms take the form given in Appendix C, the enthalpy differences are, in program language,

$$\begin{aligned} \text{DELH(I)} &= \Delta h_i \\ &= \text{BL(I)*(T3 - TX) + CL(I)*ALOG(T3/TX)} \end{aligned} \qquad \text{4-28}$$

and the fuel/air ratio is

$$\frac{F}{A} = \frac{142}{4.76\,Y\,28.97}, \qquad \text{kg/kg} \qquad \text{4-31}$$

Some notion of the size of machinery required to produce a specified power output KW can be obtained by noting that

$$\rho A V = \frac{KW}{w_{net}} \qquad \text{4-32}$$

where ρ, A, and V denote, respectively, the density, cross-sectional area, and velocity in the combustion chamber. With a large excess quantity of air in the products, the density can be simply represented by that of air, and the cross-sectional area required is then

$$A = \frac{KW}{w_{net}} \frac{R_a T_3}{V P_3} \qquad \text{4-33}$$

which can be evaluated when the velocity is specified.

If the combustion is not adiabatic, and the loss can be estimated as some fraction f of the heat of reaction, then the term H_{rp} in Eq. 4-27 would be replaced by $(1 - f) H_{rp}$.

Figures 4.4 and 4.5 illustrate some results, with zero pressure drop across the regenerator. Note that high thermal efficiency is obtained with large regenerator effectiveness and low pressure ratio across the compressor and turbine. On the other hand, large net work output will be obtained for large pressure ratios. In Fig. 4.4, observe that the lines for several values of regenerator effectiveness converge as the pressure ratio increases. Why?

Combustion chamber diameter decreases as pressure ratio increases (Fig. 4.6). Regenerator effectiveness is not a factor, because density has been based on T_3, which is fixed. Density could be calculated on T_x or on a combination of T_x and T_3, in which cases there will be some dependence on regenerator effectiveness.

Turning to the matter of the pressure drop through the regenerator, Fig. 4.7 illustrates a countercurrent heat exchanger, along with the temperature profiles of the air and combustion products. The slopes of the temperature profiles differ because of the slightly higher mass flow rate of products, as well as the higher heat capacity. A

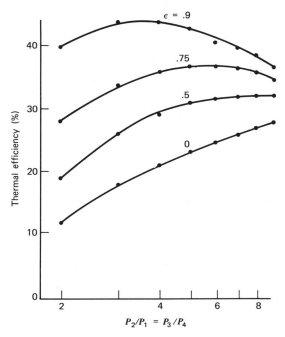

Figure 4.4 Gas turbine power plant (Fig. 4.3). The influences of pressure ratio and regenerator effectiveness on thermal efficiency. Fuel: $C_{10}H_{22}$, $T_1 = 300$ K, $T_3 = 1200$ K, $\eta_c = \eta_t = .85$, $\Delta P = 0$.

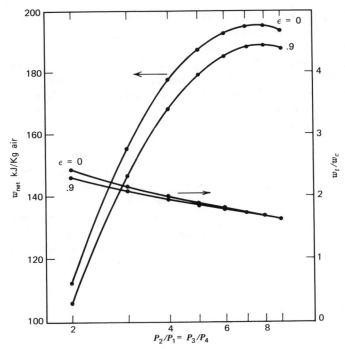

Figure 4.5 Gas turbine power plant. Influences of pressure ratio and regenerator effectiveness on w_{net} and w_t/w_c. (Same operating parameters as Fig. 4.4.)

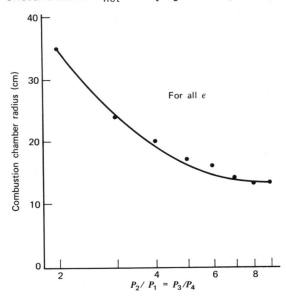

Figure 4.6 Gas turbine power plant. Influence of pressure ratio on combustion chamber radius. Power output = 746 kW, $V_{combustion\ chamber}$ = 30 m/s. (Same operating parameters as Fig. 4.4.)

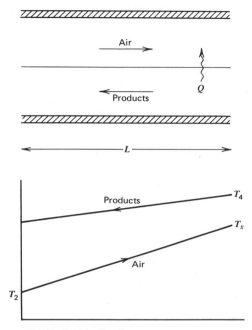

Figure 4.7 A simple countercurrent.

simple relationship between pressure drop, ΔP, and effectiveness, ϵ, can be developed by noting that as ϵ increases, so must the length, L, of the heat exchanger. We could write

$$\epsilon = 1 - e^{-L/L_o}$$

as a reasonable relation and go a step further and assume ΔP proportional to L, so that

$$\frac{\Delta P}{\Delta P_o} = -\ln(1 - \epsilon) \qquad \text{4-34}$$

Note that ΔP_o is the pressure drop for

$$\epsilon = 1 - 1/e = .63$$

With some selected value for ΔP_o, the influence of regenerator effectiveness on power plant performance will become more realistic.

Figure 4.8 illustrates a more complicated gas turbine power plant. Air compression is carried out in two stages, with interstage cooling. Two turbines are arranged in tandem, with a combustion chamber between them.

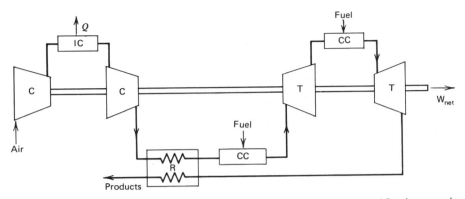

Figure 4.8 A more involved gas turbine power plant. C = compressor, IC = intercooler, T = turbine, CC = combustion chamber, R = regenerator.

4.4 AIRCRAFT GAS TURBINES

Gas turbines have been adapted for aircraft propulsion in several configurations. In the earliest configuration, (the turboprop) part of the turbine output drives the compressor, and the balance drives the propeller. Thrust is then the result of momentum change in the ambient air produced by propeller action. With the advent of higher compressor pressure ratios, the jet-prop appeared. With higher turbine outlet pressure, a substantial amount of thrust is developed from momentum change of the air passing through the engine. With still further increases in pressure ratio, the propeller could be eliminated entirely, all thrust resulting from momentum change in the air passing through the engine. That configuration is, properly speaking, a jet engine. With an increase in overall engine diameter, the fan-jet or turbofan appeared. The fan, a much reduced propeller, accelerates ambient air which passes through the annular space between the engine and a surrounding cowl (Fig. 4.9). The major portion of the engine thrust is derived from jet action. At present, aircraft gas turbines operate with compressor pressure ratios in the 10 to 25 range. Regeneration is not possible, even if it were not ruled out on account of heat exchanger bulk or the pressure drops associated with 180-degree changes in flow direction.

Since the temperature limit at the turbine inlet is reached with combustion of one-third to one-quarter of the chemically correct amount of fuel, excess air is present in substantial amount on the downstream side of the turbine. By adding fuel at that point, the temperature of the product gases entering the exhaust nozzle is increased, which augments thrust. The device that accomplishes this is known as an *afterburner* and is sketched in Fig. 4.10 and analyzed in the following discussion.

The engine moves at velocity V through air at P_o and T_o. Assuming the air that approaches the entrance to the compressor at velocity V is decelerated in a reversible,

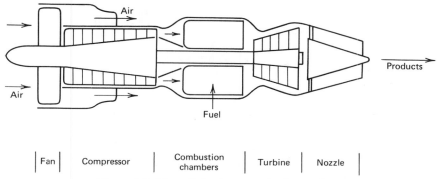

| Fan | Compressor | Combustion chambers | Turbine | Nozzle |

Figure 4.9 Aircraft fan-jet or turbofan engine.

adiabatic compression, the temperature and pressure at the compressor inlet will be

$$T_1 = T_o + \frac{V^2}{2c_{pa}}, \qquad P_1 = P_o\left(\frac{T_1}{T_o}\right)^{k_a/(k_a-1)} \qquad \text{4-35}$$

At the compressor outlet

$$T_2 = T_1\left[1 + \frac{PR_a - 1}{\eta_c}\right], \qquad PR_a = \left(\frac{P_2}{P_1}\right)^{(k_a-1)/k_a} \qquad \text{4-36}$$

and the work required to drive the compressor is

$$w_c = h_2 - h_1 = c_{pa}(T_2 - T_1) \qquad \text{4-37}$$

The energy equation for constant pressure adiabatic combustion is

$$H_r(T_2) = H_p(T_3)$$

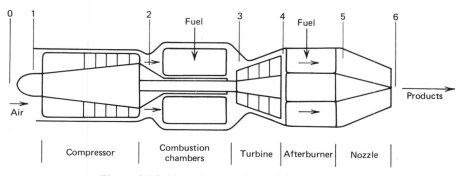

| Compressor | Combustion chambers | Turbine | Afterburner | Nozzle |

Figure 4.10 Aircraft jet engine, with afterburner.

Since additional fuel is yet to be added in the afterburner, it is convenient to represent the reactant mixture at the entrance to the combustion chamber, assuming the fuel is kerosene, as

$$X\ C_{10}H_{22} + 15.5O_2 + 58.3N_2$$

and the products are then

$$10X\ CO_2 + 11X\ H_2O + 15.5(1 - X)O_2 + 58.3N_2 \qquad \text{4-38}$$

Equation 4-37 transforms to

$$Q_p = H_p(T_3) - H_p(T_2) \qquad \text{4-39}$$

when the temperature difference, if any, between fuel and air is ignored, and now

$$Q_p = -XH_{rp} \qquad \text{4-40}$$

Given T_3, Eqs. 4-39 and 4-40 may be solved for X. Since the compressor absorbs the entire work output of the turbine,

$$w_c = w_t$$

or

$$c_{pa}(T_2 - T_1) = c_{pp}(T_3 - T_4)(1 + F/A) \qquad \text{4-41}$$

where F/A is the fuel/air ratio on a *mass* basis and occurs in the turbine work term because of the difference in mass flows through compressor and turbine. The heat capacities, c_{pa} and c_{pp}, should therefore have mass units. With

$$\frac{F}{A} = \frac{142X}{(15.5 + 58.3)28.97} \qquad \text{4-42}$$

Equation 4-41 is solved for T_4. From the definition for the turbine isentropic efficiency,

$$\eta_t = \frac{h_3 - h_4}{h_3 - h_{4s}} = \frac{T_3 - T_4}{T_3 - T_{4s}}$$

the turbine outlet pressure, P_4, can be computed,

$$P_4 = P_3 \left[1 - \frac{T_3 - T_4}{\eta_t T_3} \right]^{k_p/(k_p - 1)} \qquad \text{4-43}$$

If Y moles of fuel are burned in the afterburner, the reactant mixture will be

$$Y\,C_{10}H_{22} + 10X\,CO_2 + 11X\,H_2O + 15.5(1 - X)\,O_2 + 58.3N_2$$

resulting in a product mixture

$$10(X + Y)\,CO_2 + 11(X + Y)\,H_2O + 15.5(1 - X - Y)\,O_2 + 58.3N_2 \qquad 4\text{-}44$$

The energy equation for adiabatic constant pressure combustion is

$$Q_p = H_p(T_5) - H_p(T_4) \qquad 4\text{-}45$$

and

$$Q_p = -YH_{rp}$$

The simplest way to treat Eq. 4-45 is to solve for Y, treating T_5 as an independent variable. With no pressure loss in the afterburner,

$$P_4 = P_5$$

so that

$$T_6 = T_5 \left(\frac{P_o}{P_5}\right)^{(k_p - 1)/k_p} \qquad 4\text{-}46$$

and the velocity at the exit section of the nozzle is

$$V_6 = \sqrt{2\,[h_p(T_5) - h_p(T_6)]} \qquad 4\text{-}47$$

where h_p denotes the enthalpy of the product mixture, Eq. 4-44, in units of kilojoules per kilogram of mixture. Equation 4-47 ignores the velocity of the products in the combustion chamber.

    To evaluate the thrust, note that in time dt a mass of air dm_a is accelerated from V to V_6, and a mass of fuel dm_f is accelerated to V_6, as viewed by an observer in the engine. Then

$$\text{Thrust} = \frac{dm_a(V_6 - V) + dm_f\,V_6}{dt}$$

$$= \dot{m}_a(V_6 - V) + \dot{m}_f\,V_6$$

which may be reduced to a specific thrust, based on air flow, TH_{sp},

$$TH_{sp} = V_6\left(1 + \frac{F}{A}\right) - V, \qquad \text{N/kg of air per second} \qquad 4\text{-}48$$

or TH'_{sp}, based on fuel flow,

$$TH'_{sp} = V_6 + \frac{A}{F}(V_6 - V), \qquad \text{N/kg of fuel per second} \qquad 4\text{-}49$$

In Eqs. 4-48 and 4-49, F/A is the ratio of total fuel/air; that is,

$$\frac{F}{A} = \frac{142(X + Y)}{(15.5 + 58.3)\,28.97} \qquad\qquad 4\text{-}50$$

Some operating figures for an aircraft gas turbine equipped with an afterburner are illustrated in Table 4.1. For notation, see Fig. 4.10.

TABLE 4.1
AIRCRAFT GAS TURBINE, WITH AN AFTERBURNER
AT SEA LEVEL, STATIONARY ($V = 0$)

X	X	$X + Y$	$T5$	$T6$	Thrust		$V6$
			(K)	(K)	(N/kg air/s)	(N/kg fuel/s)	(m/s)
0.198	0.000	0.198	866	669	675	51279	666
0.198	0.075	0.273	1016	792	738	40641	725
0.198	0.163	0.361	1166	915	798	33211	779
0.198	0.269	0.466	1316	1039	856	27599	830
0.198	0.397	0.595	1466	1164	913	23108	879
0.198	0.562	0.760	1616	1294	973	19247	926
0.198	0.778	0.976	1766	1426	1034	15931	971

$P2/P1 = 15$, $T1 = 300$ K, $T3 = 1200$ K.
Compressor and turbine efficiencies = 85%.
The units for X and Y are kilomoles per 73.8 kmol of air.

With complete burning of all air, $X + Y = 1$, the thrust is increased by a little over 50%, assuming that the mass rate of air flow through the engine remains constant. Note the substantial cost in terms of fuel flow that is required for increased thrust.

In a fan-jet engine, the total thrust is

$$\text{Thrust} = [\dot{m}_a(V_6 - V) + \dot{m}_f V_6] + \rho_o S\left(V + \frac{v}{2}\right)v \qquad 4\text{-}51$$

The first term in brackets is the thrust developed by fuel and air passing through the engine; the second term is fan thrust, where S is the area of the annular region between engine and cowl, and v is the increase in axial velocity induced by fan action. (The ½ factor is the result of propeller momentum theory, which demonstrates that half of the axial velocity increase takes place on the upstream side of the propeller, or fan.) The value for v depends on fan geometry and rotating speed.

4.5 AUTOMOTIVE GAS TURBINES

The gas turbine as a power unit for automobiles and tractors has two immediate advantages: it contains fewer parts, and it can be operated on fuels that do not require extensive refining. The chief disadvantages are lower thermal efficiency than piston engines and high exhaust emissions (nitric oxide and various hydrocarbons) that may be difficult to reduce to acceptable levels (see Section 5.12). However, it seems unrealistic to attempt a meaningful comparison of gas turbines and piston engines for the simple reason that we have virtually no field experience with the former and an enormous amount of data and information about the latter. The theory of piston engine performance does not describe all that we would like to know; in many areas of performance, theory can tell us nothing, and we must resort to experiment. It seems reasonable, then, to suppose that the same need for experiment and field testing will be encountered with gas turbines. We shall confine the discussion here to a simple presentation of two proposed types of gas turbine units intended for automotive installation.

Figure 4.11 illustrates one of the features that seems to be common in all proposals; two or more turbines are used, one driving the compressor and the remainder producing work output for propulsion.

Figure 4.12 illustrates a much more complicated proposal: two compressors, four turbines, and dual combustion chambers. Air compression is staged, with intercooling; three turbines are required to perform air compression, while only one turbine produces work output to the vehicle. The proposed air compression ratio is about 16:1. Consequently, if there is to be regenerative heating of air ahead of the combustion chamber, a second combustion chamber is needed to produce high temperature in the last turbine outlet.

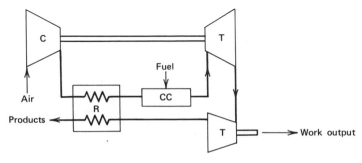

Figure 4.11 Proposed automotive gas turbine. C = compressor, T = turbine, R = regenerator, CC = combustion chamber.

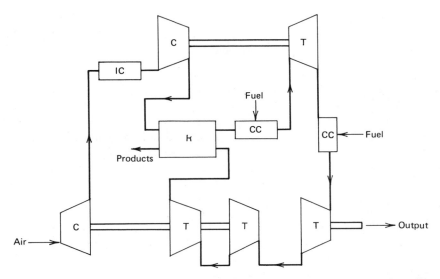

Figure 4.12 Proposed automotive gas turbine. C = compressor, T = turbine, R = regenerator, CC = combustion chamber, IC = intercooler.

4.6 COMBUSTION CHAMBERS FOR GAS TURBINES

Figure 4.13 is a sketch of what is referred to as a can-type combustion chamber. Its construction is quite simple, although the details of its operation are not, and the present state of our knowledge concerning what is taking place is very incomplete.

Air enters from the compressor, and fuel is burned. The purpose of the combustor is to send a flow of product gases to the turbine that is uniform in composition and temperature, the latter not exceeding the safe upper limit which is set by the strength of the turbine bucket material under the combined actions of high temperature and high stress resulting from high-speed rotation.

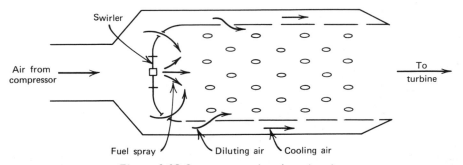

Figure 4.13 Can-type combustion chamber.

The problem is to maintain combustion in a high-speed airstream. The usual *equivalence ratio*, defined as the fuel/air ratio divided by the chemically correct fuel/air ratio, is about 0.25. This is well below the lower inflammability limit for hydrocarbons, typically 0.5. This means that even a quiescent mixture with an equivalence ratio of 0.25 would not sustain a flame once ignited, and it is more likely that such a lean mixture cannot be ignited in the first place. Continuous combustion in a gas turbine combustor is achieved by mixing sufficient air with the steady fuel spray to produce an approximately stoichiometric mixture, which will burn vigorously but reach the adiabatic flame temperature and subsequently blend in diluting air to lower the outlet temperature to the safe limit or less. This requires turbulent mixing of fuel and air in the region of the fuel spray, which is induced by a swirler installed at the head end of the combustor. This involves only a fraction of the total air flow; the remainder passes along the outside of the can, cooling the metal until it passes into the interior, where it mixes with hot combustion gases.

The problems associated with the development of a satisfactory combustor are formidable. The air and fuel mix in a turbulent pattern, which is not yet understood. The fuel spray must vaporize before it can burn. Once ignited to produce a flame front, that front will move, in a complicated pattern, into regions not yet ignited and cause burning in the new region, provided that the fresh mixture is ready to burn and that the flame front itself does not burn out, that is, go to completion. The kinetics of hydrocarbon combustion, which is concerned with the time rate at which reactants pass over into products with unchanging composition, is not yet understood. All this means that combustors evolve by trial-and-error techniques, aided by observation of events in the chamber.

For aircraft gas turbine units, it is imperative that the overall diameter of the combustion section, which comprises a group of nested can-type combustors, be kept as small as possible in order to avoid excessive drag.

4.7 COMBINED STEAM TURBINE – GAS TURBINE PLANTS

Gas turbine exhaust may be passed through a heat exchanger to generate steam. The steam then produces power in addition to the net output from the gas turbine-compressor combination, or the steam may then produce all of the net output to the surroundings. Figure 4.14 illustrates the heat exchanger. The problem is to evaluate the amount of steam that can be produced per unit mass of exhaust. The catch to the solution lurks inside the exchanger and can be understood at once by drawing the temperature profiles of the two streams. Presumably, the water enters the exchanger in the liquid state, in which case the H_2O will exhibit the constant temperature plateau that occurs during the phase change. The temperature difference between entering exhaust and leaving steam is referred to as the *superheat* ΔT, while the temperature difference between exhaust and water at the point in the heat exchanger where water reaches the saturated liquid state is referred to as the *pinch point* ΔT. To avoid excessively long exchangers, nowhere should the temperature

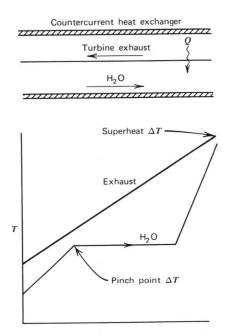

Figure 4.14 Heat exchange between gas turbine exhaust and water.

difference between the two streams drop below about 30°C. Given the pressure on the water side of the exchanger, the exhaust temperature at the pinch point can be assigned a known minimum. The entering exhaust temperature fixes the maximum allowable steam exit temperature. In addition, for purely practical reasons, it is necessary to examine the condition of the exhaust leaving the exchanger. Sulfuric acid condensation in the exchanger or stack produces unwanted corrosion, which places a lower temperature limit on the exhaust at somewhere between 120°C for clean fuels, and 180°C for high sulfur grade fuels. On the other hand, if the fuel is free of sulfur, condensation of water vapor in the exhaust should be avoided, and the corresponding minimum exit temperature is fixed by the water vapor mole fraction in the exhaust gas mixture.

When the superheat and pinch point ΔT's are selected, the maximum enthalpy rise of the water between saturated liquid (at the pinch point) and exit temperature is fixed. It is then necessary to compute the exit temperature of the exhaust leaving the heat exchanger. When it falls below the minimum allowable to avoid condensation, the superheat and pinch point ΔT's must be raised.

Figure 4.15 illustrates a simple combination of gas and steam turbines. The heat exchanger in Fig. 4.14 acquires a new identification and is known as the *heat recovery*

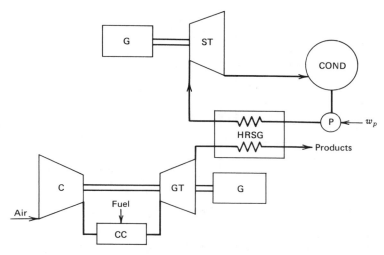

Figure 4.15 Combined steam and gas turbine power plant. C = compressor, CC = combustion chamber, GT = gas turbine, ST = steam turbine, COND = condenser, P = pump, G= generator, HRSG = heat recovery steam generator.

steam generator. The feature to note in Fig. 4.15 is the increase in work output to the surroundings with no additional supply of fuel. Now recall the regenerative gas turbine cycle shown in Fig. 4.3, in which fuel supply is reduced, with no decrease in work output to the surroundings. Consequently, Figs. 4.3 and 4.15 illustrate opposite approaches to the problem of increasing thermal efficiency. A proper thermodynamic analysis should, however, include the pressure losses associated with the heat exchangers, whether they be regenerators or steam generators.

There is one essential difference between the regenerative power plant and the combined turbine power plant. The regenerative power plant is limited to a maximum compressor pressure ratio (Eq. 4-16), while the combined turbine power plant is not. This difference sets the two power plants apart; it also limits the extent to which they may be compared.

Figure 4.16 illustrates another form of combined steam and gas turbine power plant, this one involving the gasification of coal, which is sustained by the injection of high temperature air from the second-stage compressor. After the coal gas has been cleaned, it fires the combustion chamber ahead of the gas turbine, and the gas turbine exhaust, together with intercooling between air compressors, produces steam to drive the steam turbine. This is one of the more advanced forms of steam and gas turbine integration. A good deal of study and effort centers on the operation of the gasifier and cleaning operations. The concept of coal gasification appears to have considerable promise.

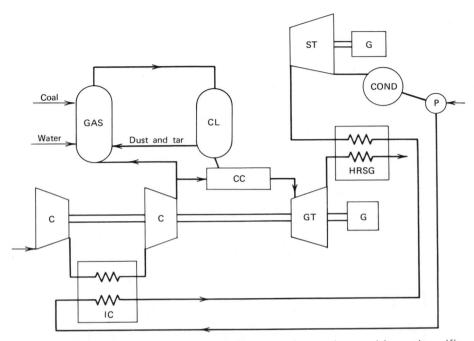

Figure 4.16 Combined steam and gas turbine power plant, with coal gasifier. C = compressor, CC = combustion chamber, GT = gas turbine, P = pump, St = steam turbine, Gas = gasifier, CL = cleaner, G = generator IC = intercooler, HRSG = heat recovery steam generator.

REFERENCES

When the analysis of gas turbine cycle performance is to be carried beyond the limited contributions that are possible from thermodynamics, the next topic to be considered is the burning rate of fuels, that is, the kinetics of reactions. A recommended text that contains an excellent introduction to kinetics is:

A. Murty Kanury, *Introduction to Combustion Phenomena*, Gordon & Breach Science Publishers, New York, 1975.

Two journals, the *Transactions of the Society of Automotive Engineers* and *Combustion Science and Technology*, report recent experimental and theoretical information on gas turbine performance.

PROBLEMS

4.1 For a stationary gas turbine unit, such as that shown in Fig. 4.1*b*, calculate T_2, T_4, thermal efficiency, and work ratio for

$$P_2/P_1 = 5, 10, \text{ and } 15$$

with

$$T_3 = 1300 \text{ K}, \qquad T_1 = 300 \text{ K}, \qquad c_p = 30 \text{ kJ/kmol K}$$

Observe how various features of the air standard Brayton cycle vary with pressure ratio. Graphs drawn with compressor and turbine efficiency as the abcissa may be helpful. Various values for these efficiencies are to be selected.

    4.2   Air enters a combustion chamber at 500 K under steady flow. Liquid kerosene, $C_{10}H_{22}$, is injected in sufficient quantity to produce a temperature of 1300 K in the products of combustion. What fuel/air ratio (kg/kg) is required?

    4.3   The rate of fuel flow into the combustion chamber of Problem 4.2 could be controlled by some sort of metering device. That device, no matter how carefully it has been designed and built, will operate with some tolerance. If the tolerance is ±4%, what will be the maximum and minimum temperatures leaving the combustion chamber?

    4.4   The product gases in Problem 4.2 approach a turbine at a stream temperature of 1300 K and a velocity of 170 m/s. The turbine is 30 cm in diameter and turns at 26,000 rpm. What is the stagnation temperature at the tip of the turbine wheel?

    4.5   For a gas turbine power plant fueled with kerosene and equipped with a regenerator (Fig. 4.3), write a computer program that will evaluate

    Thermal efficiency
    Net work
    Fuel/air ratio
    Work ratio
    Combustion chamber radius

for given values of

    Compressor inlet temperature
    Turbine inlet temperature
    Turbine and compressor efficiencies
    Regenerator effectiveness
    Compressor pressure ratio
    Turbine pressure ratio
    Net power output
    Combustion chamber gas velocity

Such a program can be checked against Figs. 4.4, 4.5, and 4.6 and can be used to examine a number of features of power plant behavior and plant size.

    4.6   Two turbines are to be operated in tandem, as illustrated in Fig. 4.17. Turbine I has an efficiency of 1.0, while the efficiency of turbine II is less than 1.0.
    In both arrangements the inlet conditions P_1 and T_1 are identical. Also,

$$P_1/P_2 = P_2/P_3$$

Which arrangement will produce the largest power output?

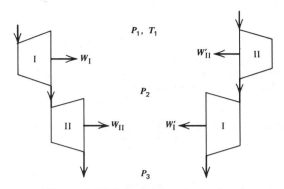

Figure 4.17 Two turbines in tandem

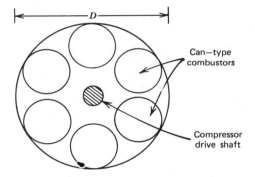

Figure 4.18 Combustion chamber cross-section showing six nested combustors.

4.7   For what value of P_2/P_1 will *all* the curves in Fig. 4.4 meet together?

4.8   The cross section of the combustion chamber of aircraft gas turbines usually appears, as shown in Fig. 4.18, as a set of equal diameter can-type combustors nested together within an overall diameter D. The engine can be designed with n combustors.

If D is fixed, what factors might be considered in choosing a value for n?

4.9   An aircraft gas turbine, equipped with an afterburner, has a combustion chamber as shown in Fig. 4.18. The overall diameter D is 1 m.

What do you need to know in the way of operating conditions in order to calculate the thrust the engine will deliver? Specify those conditions, and calculate the thrust in newtons or kilonewtons.

5

FOUR-STROKE INTERNAL COMBUSTION ENGINES

5.1 INTRODUCTION

This chapter describes the techniques that produce the analysis that leads to thermal efficiency and power output calculations for Otto and Diesel cycle four-stroke engines, both rotary and reciprocating piston designs.

It may appear that more attention and emphasis is devoted to the Otto cycle and less to the Diesel. In fact, the two cycles share a number of common features. Many aspects of the analysis of one cycle apply, without alteration or with only minor change, to the other.

The material is presented in the manner that adapts most easily to the construction of a computer program to carry out the calculations. Hand computations are tedious. Also, the computations must begin with some assumptions as to the state of affairs in the engine, and enough cycles run through to the point where the numbers cease to change between successive cycles. And lastly, of course, the number of parameters that can be specified independently is quite large. The list can include the following:

Engine speed
Fuel, or mixture of fuels
Intake manifold pressure and temperature
Exhaust manifold pressure
Fuel/air ratio
Compression ratio
Type of supercharging
Spark advance
Combustion chamber geometry
Spark plug location

Rate of fuel injection
Size of exhaust and intake valve openings
Valve cam profile
Valve timing
Engine size
Heat transfer coefficients

The objective of this chapter is to demonstrate how and why these operating conditions and design features affect engine performance. The primary tool of the analysis is thermodynamics, but as we shall see, thermodynamics alone will not carry us very far. To deal with the combustion process, for example, we must resort to experimental results and empirical formulations. Even with these, it is evident that our comprehension of what is taking place during the burning process can only be described as rudimentary.

The published literature dealing with internal combustion engine performance is enormous. Some references are listed at the end of the chapter.

5.2 ENGINE KINEMATICS

Four-stroke internal combustion engines are currently produced in two configurations, reciprocating piston and rotary.

The basic geometry of the reciprocating piston engine is shown in Fig. 5.1. It is described in terms of cylinder bore (i.e., diameter) B, length of stroke S, length of connecting rod L, and compression ratio CR. The displacement volume, VDISP, is swept out as the piston moves from bottom dead center (BDC), or extreme outer position, to top dead center (TDC) or extreme inner position. Thus

$$\text{VDISP} = \text{VBDC} - \text{VTDC} = \frac{\pi}{4} S(B^2) \qquad \text{5-1}$$

and the compression ratio is

$$CR = \text{VBDC/VTDC} \qquad \text{5-2}$$

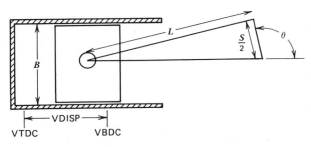

Figure 5.1 Geometry of reciprocating piston engine.

so that we may also write

$$\text{VBDC} = \frac{CR}{CR - 1} \text{ VDISP}, \quad \text{VTDC} = \frac{1}{CR - 1} \text{ VDISP} \qquad 5\text{-}3$$

With θ denoting the angular displacement of the crank from BDC, the volume V is represented by

$$V(\theta) = \text{VDISP} \left[\frac{CR}{CR - 1} - \frac{1 - \cos \theta}{2} + \frac{L}{S} - \frac{1}{2} \sqrt{\left(2 \frac{L}{S} \right)^2 - \sin^2 \theta} \right] \qquad 5\text{-}4$$

and

$$V'(\theta) = \frac{dV}{d\theta} = \frac{\text{VDISP}}{2} \left[\frac{1}{2} \frac{\sin 2\theta}{\sqrt{\left(2 \frac{L}{S} \right)^2 - \sin^2 \theta}} - \sin \theta \right] \qquad 5\text{-}5$$

Displacement volume is an important measure of engine design, as it is directly related to theoretical work output in an ideal engine cycle.

Figure 5.2 illustrates a rotary engine, one among the various possible designs; in this case, a planetary rotor configuration is shown. A *rotor* in the shape of an equilateral triangle with curved flanks between the apexes turns in a *bore*, the enclosing member of which is generally oval in shape, with slight necking-in at the minor axis. The dark circle at the center denotes the axis of rotation of the power shaft. To that shaft is fastened an eccentric, and the rotor rides on that eccentric. Seals at the three apexes, A, B, and C, are the rotary counterparts of piston rings. Planetary motion is achieved through an arrangement of gears. The rotor contains a ring gear that meshes with a smaller spur pinion, the latter being fixed to one of the end covers, which, together with the bore, define the three volumes formed by the rotor. Thus, in Fig. 5.2, the rotor performs planetary motion, and as shown, the three volumes change in size as the rotor moves. (When viewed for the first time, the sense of astonishment is not so much that the rotary engine works well but rather that it works at all!)

Depressions can be machined into the face of each rotor flank, as indicated in Fig. 5.2 by the dashed lines. These depressions increase the working volume. They are plainly shown in the view of a twin rotor engine in Fig. 5.3

The relation between working volume and rotor angle is

$$V(\alpha) = \frac{\text{VDISP}}{2} \left[\frac{CR + 1}{CR - 1} + \cos 2\alpha \right] \qquad 5\text{-}6$$

rotor angle α being measured from the position of maximum volume. Maximum and minimum volumes in a rotary engine correspond to VBDC and VTDC for a reciprocating piston engine.

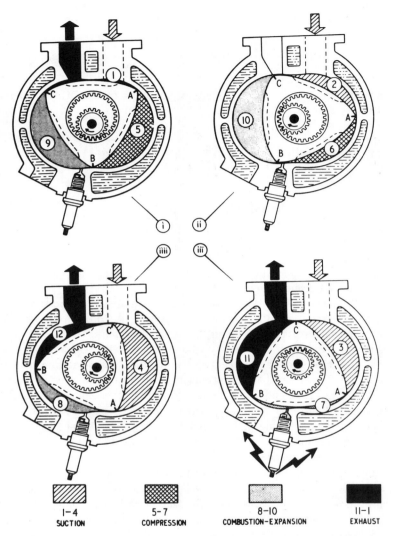

1-4 SUCTION	5-7 COMPRESSION	8-10 COMBUSTION-EXPANSION	11-1 EXHAUST

Figure 5.2 A planetary rotary engine (Reprinted from "The Wankel RC Engine" by R. F. Ansdale, A. S. Barnes & Co., New York and Butterworth & Co., Ltd, London, with permission.)

The rotary design offers a particularly fascinating variety of design options, and the engine appears to hold unusual future promise. Problems have been encountered in the short time since its development began. In assessing the future of the rotary engine, it should be remembered that the present-day reciprocating engine is the product of almost 100 years of intensive effort by large scientific and technical staffs.

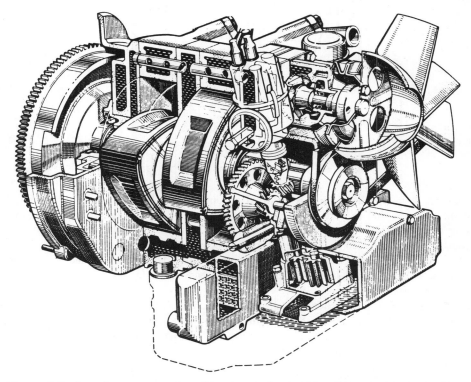

Figure 5.3 A twin rotor engine (Reprinted from "The Wankel RC Engine" by R. F. Ansdale, A. S. Barnes & Co., New York and Butterworth & Co., Ltd, London, with permission.)

There are several fundamental differences between reciprocating piston and rotary engines. These will be described in appropriate sections later in this chapter.

5.3 CARBURETION AND FUEL INJECTION

In spark ignition engines of whatever design, the cylinder contains fuel vapor, air, and residual exhaust from the previous cycle at the instant a spark passes between the electrodes. The fuel may be introduced in three ways:

1. Carburetion
2. Intake manifold injection
3. Cylinder head injection

Whichever system is employed, each must incorporate a device that meters fuel flow in accordance with air flow. The general form of the relation between the two flows for an automotive engine is shown in Fig. 5.4. At idle conditions, where power and air

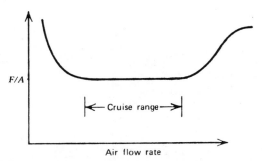

Figure 5.4 An example of fuel and air flow requirements.

flow are low, a rich mixture is required because of the large dilution from residual exhaust gases, due in part to low speed and in part to valve overlap. At high power output, where air flow is large, a rich mixture is needed. In the cruise range, the fuel/air ratio is reduced in the interests of fuel economy.

In a carburetor, fuel and air flows are coordinated, and fuel is injected into the airstream. In manifold injection and cylinder head injection, the two processes are separated from each other.

Figure 5.5 is a schematic diagram of some of the features of automotive carburetors. As the air passes through the venturi throat, air pressure is reduced, and fuel is drawn through the main metering jet from the fuel bowl and enters the airstream, where it vaporizes. The air bleed tends to reduce the amount of suction causing fuel flow, and the combination of air bleed and main metering jet will generally produce a rather flat fuel/air curve throughout the range of air flow. At idle, where the throttle plate closes almost completely, a high vacuum is present on the downstream side. Practically no fuel is drawn through the main metering jet; instead, it is drawn through the idle jet and passes through an adjustable restriction. At high power output, which requires a wide-open throttle setting, additional metering jets are gradually uncovered by way of a linkage connected to the throttle shaft (not shown in Fig. 5.5). The fuel bowl is vented to air on the upstream side of the venturi to compensate for changes in ambient air pressure. The float actuates a needle which closes the inlet valve when the level of fuel in the bowl rises to some preset level. Rapid opening of the throttle calls for a correspondingly rapid increase in fuel flow. This is achieved by a plunger linked to the throttle shaft and moving in a cylinder filled with fuel (not shown in Fig. 5.5). With a suitably sized bypass jet, the accelerating pump is virtually inoperative as a supplier of fuel during gradual movement of the throttle, but it is effective for sudden acceleration. Automotive carburetors are superb examples of engineering development. The principles upon which they operate are simple and easy to grasp: pressure drop induced by air flow, pressure drop inducing fuel flow. But the production, in a small package, of a device that is both reliable and trouble free, and at the same time manufactured to conform to close tolerance specifications, requires imagination and considerable experimentation.

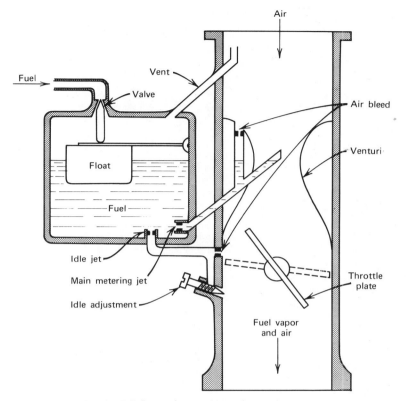

Figure 5.5 Part of an automotive carburetor.

Small engines, below, say, 10 hp or 7 kW, are not equipped with such sophisticated carburetors. In general, the smaller the engine, the simpler the carburetor. Very
small engines reduce the whole system to a gravity feed, a fuel bowl and float-actuated
valve, and a single adjustable metering jet. On the other hand, small engines must, in
some applications, be capable of continuous operation when turned upside down. The
carburetor in Fig. 5.5 would fail that test.

In the case of aircraft carburetors, some provisions must be incorporated for the
changes in ambient air pressure and temperature that occur with altitude. This can be
accomplished, referring to Fig. 5.5, by drawing a continuous, but small, rate of air
flow through the air bleed and fuel bowl, discharging the flow below the throttle.
Then by controlling the rate of that flow with, say, a bellows-operated restrictor in the
air bleed, a crude altitude compensator is achieved. Aircraft carburetors must also
tolerate large changes in attitude, going as far as upside down.

Figure 5.6 illustrates two types of fuel injection, with a low-pressure nozzle
located in the intake pipe ahead of the intake valve and a high-pressure injection

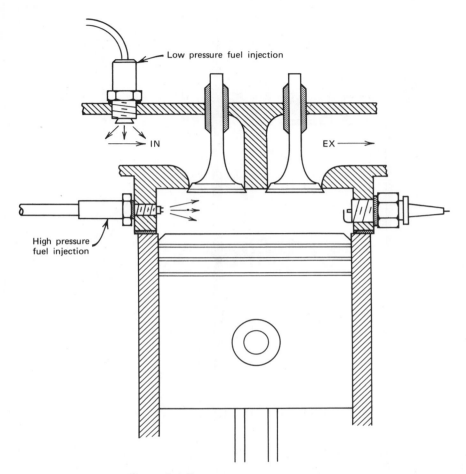

Figure 5.6 Two forms of fuel injection.

nozzle located in the cylinder head. In its simplest form, the low-pressure manifold injection system meters fuel pressure in accordance with air flow, and the fuel flow can be continuous. In more involved systems, the flow is pulsed. Usually, provision must be incorporated for abundant fuel flow for sudden throttle openings.

Cylinder head injection is vastly more complicated. To begin with, the flow of fuel must be synchronized with piston motion and with spark advance. Ideally, all the fuel is in the cylinder and vaporized before the spark passes. This requires a fine spray, which in turn calls for high line pressure, 100 atm and up. Figure 5.7 illustrates one form of fuel pump, supplying six cylinders. A separate plunger feeds each cylinder. The plungers are attached for to-and-fro motion to a rotating swash plate or wobble

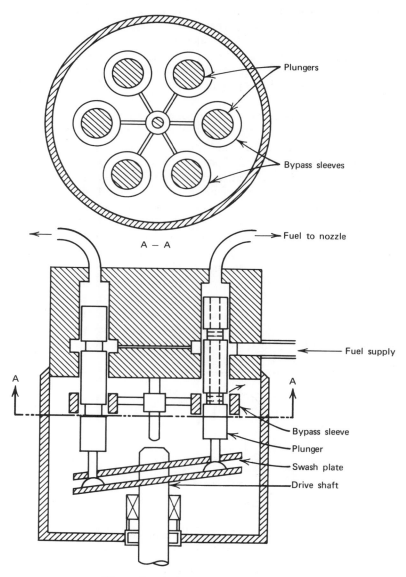

Figure 5.7 A form of injection pump.

plate. Each plunger is machined with two lands, the upper land providing the channel for filling the plunger space and the lower for metering the quantity of fuel pumped, which is accomplished by movable bypass sleeves. The bypass sleeves are ganged together for uniform fuel flow to each cylinder and are positioned through mechanical linkage to a device that monitors air flow. The technology of fuel injection pumps has borrowed heavily on experience with Diesel injection. Pumping gasoline, however, is in some ways not as simple as pumping an oil.

From the point of view of thermodynamics, the manner in which fuel is introduced is of no consequence for work output calculations. All methods will lead to the same numbers or very close, say 1%, to the same numbers. Why, then, go to the expense and bother of fuel injection? The answer lies in the need, or desire, to secure a uniform fuel/air ratio in all cylinders. Consider, for example, the four cylinder in-line engine block, pictured in Fig. 5.8, with a carburetor and intake manifold shown. Now it is not unreasonably difficult to form the shapes and sizes of the manifold pipes to produce equal air flows through the cylinders. But equal fuel flow is an altogether different matter, largely because the fuel dribbles into the airstream, breaks up into small drops, and then evaporates in a manner that is not under our control. Furthermore, the time available for evaporation shortens as speed increases. Fuel drops, which are slow to evaporate, will not follow airstream lines, and as a consequence, large discrepencies in the fuel/air ratio can develop. If, for example, the fuel/air ratio is set for best power or for best economy, with poor fuel distribution, each cylinder may be operating above or below the optimum, with a loss in total engine performance. Besides uniform fuel distribution, fuel injection systems are safer with fuel confined to the cylinder or very close to it. And as we shall see in the following discussion,

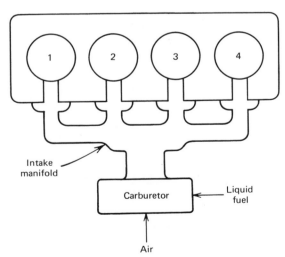

Figure 5.8 Four-cylinder engine with carburetor and intake manifold.

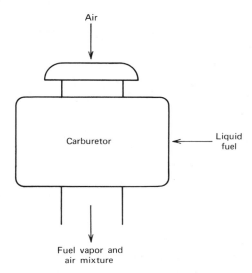

Figure 5.9 A carburetor, viewed as a thermodynamic process.

vaporization causes a temperature drop in the airstream, with the possibility of ice buildup below the throttle, which can lead to a number of unwanted conditions.

Referring to Fig. 5.9, suppose NA moles of air enter the carburetor at temperature TA, while 1 mole of liquid fuel enters at TA. Both flows are steady. Ignoring heat transfer to the surroundings,

$$NAh_a(TA) + h_f(\text{liq}, TA) = NAh_a(T) + h_f(\text{vap}, T)$$

where T is the temperature downstream after vaporization, and a and f subscripts denote air and fuel. Adding and subtracting the term

$$h_f(\text{vap}, TA)$$

on the right-hand side of the equation leads to

$$\Delta T = T - TA$$

$$= \text{temperature change due to vaporization}$$

$$\Delta T = \frac{h_{fg}}{(A/F)CPA + CPF} \qquad \text{5-7}$$

where CPA and CPF denote the specific heats of air and fuel, h_{fg} the heat of vaporization, and A/F the air/fuel ratio. Table 5.1 shows the order of magnitude for

TABLE 5.1
TEMPERATURE DROPS CAUSED BY VAPORIZATION

$TA = TF = 300$ K; $CPA = 29$ kJ/kmol K

Fuel	$(F/A)_{cc}$ (mol/mol)	h_{fg} (kJ/kmol)	CPF (kJ/kmol K)	ΔT (°C)
C_8H_{18}	1/49.5	41 300	167	25
C_6H_6	1/35.7	33 900	93	30
C_3HOH	1/7.1	37 380	46	148
C_2H_5OH	1/14.3	42 200	67	88

ΔT for several fuels, with chemically correct F/A ratio. Values for h_{fg} are calculated from Table E.4, while the fuel heat capacities are calculated from Table E.1. Methanol (C_3HOH or methyl alcohol) is mentioned as a partial substitute for gasoline, C_8H_{18}, and the difference in the ΔT's indicates that large temperature changes in the airstream may result unless provisions are incorporated to supply heat on the downstream side of the carburetor.

Equation 5-7 would apply without change to the case of intake manifold injection. When injection is made directly into the combustion chamber, matters are treated by a different scheme, which is discussed in Section 5.6.

5.4 AIR STANDARD OTTO CYCLE

Figure 5.10 illustrates the Otto cycle. The figure is drawn approximately to scale — 9:1 compression ratio, a 4.3 pressure ratio during combustion, and a specific heat ratio of 1.4. Isentropic compression occurs between 1 and 2, followed by constant volume adiabatic combustion from 2 to 3, isentropic expansion from 3 to 4, and constant volume cooling from 4 back to 1. However, in the air standard cycle the combustion process is replaced by heating from an exterior source. We are interested in the thermal efficiency, which involves the net work per unit quantity of air

$$w_{net} = c_v(T_3 - T_2) - c_v(T_4 - T_1)$$

and the heat added,

$$q_a = c_v(T_3 - T_2)$$

so the thermal efficiency, η_{th}, is

$$\eta_{th} = 1 - \frac{T_1(T_4/T_1 - 1)}{T_2(T_3/T_2 - 1)}$$

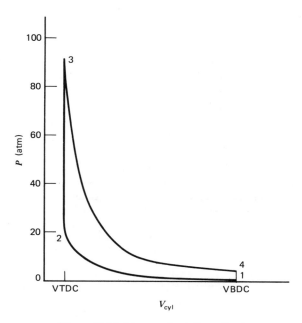

Figure 5.10 Air standard Otto Cycle

but

$$\frac{T_2}{T_1} = \frac{T_3}{T_4} = CR^{k-1}, \qquad \text{hence } \frac{T_4}{T_1} = \frac{T_3}{T_2}$$

and the thermal efficiency reduces to

$$\eta_{th} = 1 - \frac{1}{CR^{k-1}} \qquad\qquad 5\text{-}8$$

5.5 IDEAL OTTO CYCLE: FULL THROTTLE

Figure 5.11 illustrates the ideal Otto cycle. It differs from the air standard cycle (Fig. 5.10) by the inclusion of two additional processes, the exhaust stroke from 1 to 5 and the intake stroke from 5 to 1. Figure 5.11 is deliberately distorted to show these two strokes. Furthermore, in this analysis of the ideal cycle the constant volume process from 2 to 3 will be brought about by combustion.

The analysis that follows in this section and throughout this chapter will

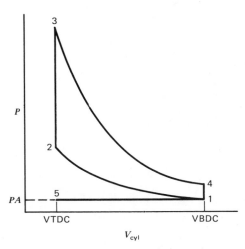

Figure 5.11 The ideal Otto cycle.

examine the performance of a piston/cylinder engine with the following basic geo-metry:

$$B = \text{bore} = 10.16 \text{ cm} = 4.0 \text{ in.}$$
$$S = \text{stroke} = B, L = \text{connecting rod length} = 2B$$
$$CR = \text{compression ratio} = 9{:}1$$
$$\text{VDISP} = 823 \text{ cm}^3 = .823 \text{ liter} = 50.27 \text{ in}^3.$$
$$\text{VTDC} = 103 \text{ cm}^3, \text{VBDC} = 926 \text{ cm}^3$$

Alternatively, the analysis will apply to a rotary engine with the same VDISP, VTDC, and VBDC for each rotor flank.

For a given fuel, fuel/air ratio, engine speed, and ambient air temperature, we wish to calculate the power output. Engine speed would be crankshaft revolutions per minute (rpm) for the reciprocating piston engine, and rotor rpm for the rotary engine. The calculated power output will then be associated with one cylinder of a multi-cylinder engine or with one flank of a multiflank rotor. In this section, we consider full throttle operation, that is

$$P1 = P5 = PA = \text{ambient pressure}$$

In the next section, the analysis is enlarged to include operation at manifold pressure below ambient.

Let TA = ambient air temperature and TM = intake manifold temperature. Then

$$TM = TA - \Delta T \qquad\qquad 5\text{-}9$$

where ΔT is given by Eq. 5-7. If the fuel arrives at the carburetor already in vapor form, as with methane or propane for example, then, of course, $\Delta T = 0$. The following definitions prove useful in the analysis:

> NMO = kilomoles of air plus fuel vapor in a mixture containing 1 kmol of fuel.
> NPO = kilomoles of products formed from the combustion of NMO.
> NM = kilomoles of fuel vapor and air in the engine during the compression stroke
> NX = kilomoles of residual exhaust from previous cycle in the engine during compression stroke.
> NP = kilomoles of products formed from the combustion of $NM + NX$.

The calculations chase a mixture of fuel vapor and air as they leave the carburetor and pass through the engine. Obviously, we are not dealing with a cycle, at least not for the working fluid, since it changes chemical composition and is not returned to the original state of fuel and air. On the other hand, the operation of the engine is cyclic. We commence at the start of compression, and we have to assume values for NX and for $T1$; both depend on the previous cycle, which depends, in turn, on the cycle before that, and so on. Let us set

$$NX = 0 \qquad \text{and} \qquad T1 = TM \qquad\qquad 5\text{-}10$$

(The calculations will converge for any values of NX and $T1$, even wildly absurd values. The proof for convergence lies in the observation that real engines run steadily.) Then

$$NM = \frac{P1\ VBDC}{R\ TM} \qquad\qquad 5\text{-}11$$

Let CPR = constant pressure heat capacity (kJ/kmol K) for the mixture $NM + NX$, and

$$KR = CPR/(CPR - 8.314).$$

Then

$$T2 = T1\ CR^{KR\ -\ 1}, \qquad P2 = P1\ CR^{KR} \qquad\qquad 5\text{-}12$$

and

$WCOMP$ = work done on the reactants, $NM + NX$, by the rotor or piston during adiabatic compression

$$= (NM + NX)(CPR - 8.314)(T2 - T1) \qquad\qquad 5\text{-}13$$

The combustion is assumed to occur adiabatically and instantaneously, with the rotor or piston stationary; hence the energy equation will read

$$U_r(T2) = U_p(T3) \qquad\qquad 5\text{-}14$$

which will lead to a value for $T3$, and hence to

$$P3 = P2 \frac{T3}{T2} \frac{NP}{NM + NX} \tag{5-15}$$

We shall return to examine Eq. 5-14 in more detail. Now let

     CPP = constant pressure heat capacity (kJ/kmol K) for the mixture NP, and

     $KP = CPP/(CPP - 8.314)$.

Then

$$T4 = \frac{T3}{CR^{KP-1}} , \qquad P4 = \frac{P3}{CR^{KP}} \tag{5-16}$$

and

$WEXP$ = work done by the products NP on the rotor or piston during adiabatic expansion

$$= U_p(T3) - U_p(T4) \tag{5-17}$$

and the net work output for the cycle will be

$$\text{WNET} = \text{WEXP} - \text{WCOMP} \tag{5-18}$$

At point 4 we assume that the exhaust valve opens, that the pressure drops to ambient, and that the temperature of the exhaust drops to

$$T5 = T4 \left(\frac{P1}{P4}\right)^{(KP-1)/KP} \tag{5-19}$$

As the engine moves from BDC to TDC, most of the exhaust is pushed out; what remains in the engine at TDC is

$$NX = \frac{P1 \text{ VTDC}}{R \text{ } T5} \tag{5-20}$$

The exhaust valve now closes, the intake valve opens, and the piston or rotor moves slowly to BDC position, during which time NM moles of fuel/air mixture are drawn in to the working space; after mixing with NX the resulting temperature will be $T1$. Note that NM is *not* given by Eq. 5-11; NM and $T1$ are to be found by analyzing the intake process, shown in Fig. 5.12.

     The fuel/air mixture that will be drawn into the engine during the intake stroke initially occupies volume VM in the intake pipe. Place a control mass around

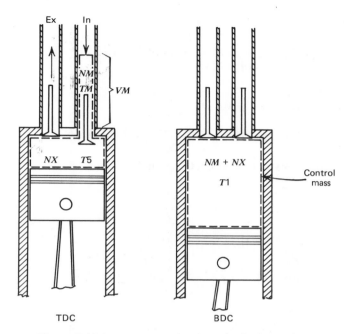

Figure 5.12 Intake stroke in piston/cylinder engine.

$NM + NX$, as shown in Fig. 5.12. The initial energy in the control mass is

$$NX u_p(T5) + NM u_m(TM) \qquad \text{5-21}$$

As the piston moves through the length of the stroke, work amounting to

$$P1(\text{VBDC} - \text{VTDC}) \qquad \text{5-22}$$

will be done *by* the gases in the working space, while work amounting to

$$P1 \, VM \qquad \text{5-23}$$

is done *on* the control mass by the fuel/air mixture behind NM. At BDC the energy in the control mass is

$$NX u_p(T1) + NM u_m(T1) \qquad \text{5-24}$$

These four terms combine in an energy equation;

$$NX\, u_p(T5) + NM\, u_m(TM) - P1(\text{VBDC} - \text{VTDC}) + P1\ VM =$$
$$NX\, u_p(T1) + NM\, u_m(T1) \tag{5-25}$$

and by grouping U and PV terms, we get

$$NX\, h_p(T5) + NM\, h_m(TM) = NX\, h_p(T1) + NM\, h_m(T) \tag{5-26}$$

Replacing enthalpy terms by $c_p T$ products, and noting that

$$NM = \frac{P1\ \text{VBDC}}{R\ T1} - NX \tag{5-27}$$

the energy equation reduces to

$$\left\{\frac{CPP}{CPM} - 1\right\} \frac{T1^2}{TM\ T5} + \left\{CR + \frac{TM}{T5} - \frac{CPP}{CPM}\right\} \frac{T1}{TM} - CR = 0 \tag{5-28}$$

where

CPM = constant pressure heat capacity (kJ/kmol K) of the fuel/air mixture.

Equation 5-28 may be solved for $T1$. If the difference between CPP and CPM is ignored, Eq. 5-28 reduces to a simpler expression,

$$T1 = \frac{CR\ TM}{CR - 1 + TM/T5} \tag{5-29}$$

The intake process in a rotary planetary engine is illustrated in Fig. 5.13. The rotor is moving in the clockwise direction. One apex is about to begin to cover the intake part. (An important advantage for rotary engines is the absence of valves and the associated mechanisms which, as is well known, tend to limit the capacity of piston/cylinder engines to induct air. The design options associated with the location and shape of port openings in rotary engines are many and varied.) Note that the resultant force of pressure acting on the right-hand flank, P, passes through the center of gravity of the symmetrical rotor but not through the center of the power shaft. The offset, which is the result of the eccentric mounting of the rotor, produces torque.

All the equations leading up to Eqs. 5-28 and 5-29 apply to rotary engines; internal energy is independent of shape, and the two work expressions are derived from $P\,dV$ integrals, which are also independent of geometry.

To return to the chain of calculations, with new values for $T1$, NM, and NX, a new value for CPR is calculated, and the computations proceed through a second

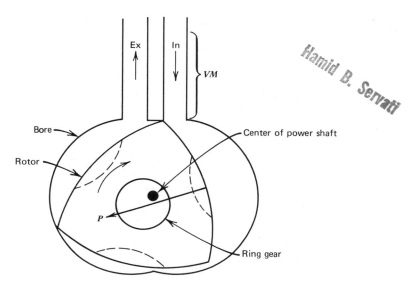

Figure 5.13 Intake stroke in a planetary rotary engine.

cycle, beginning with the compression stroke, Eqs. 5-12. A third cycle can follow the second, and so on. Clearly, the computations are to be halted when successive cycles produce the same numbers. Generally, the convergence is rapid.

We now return to the energy equation, 5.14, which can be rewritten

$$Q_v = \Sigma\{N(I)[u_I(T3) - u_I(T2)]\} \qquad\qquad 5\text{-}30$$

which has the same form as Eq. 2-31 and can be conveniently solved by a Newton-Raphson iteration once we have appropriate values for the product gas mole numbers, the $N(I)$ in Eq. 5-30. We will evaluate these mole numbers in terms of the actual values for the engine under consideration. In doing so, it is perhaps easier for the reader to follow a specific example; the results transfer simply for any other choice of fuel.

Suppose the fuel is C_8H_{18}, a common substitute for gasoline in thermodynamic analysis. Let the correspondence between I values and product gases be

$$1 = CO, \qquad 2 = CO_2, \qquad 3 = H_2O, \qquad 4 = N_2, \qquad 5 = O_2$$

With Y representing moles O_2/mole C_8H_{18}, the fuel/air mixture is

$$C_8H_{18} + Y O_2 + 3.76Y N_2 \qquad\qquad 5\text{-}31$$

and therefore, from the definitions given above,

$$NMO = 1 + 4.76Y$$
$$NPO = 17 + 3.76Y, \qquad \text{for } 8.5 \leqslant Y \leqslant 12.5 \qquad\qquad 5\text{-}32$$
$$NPO = 4.5 + 4.76Y, \qquad \text{for } Y \geqslant 12.5$$

The heat of reaction for gaseous octane, Table E.4, is

$$H_{rp} = -5\,089\,100 \text{ kJ/kmol } C_8H_{18} \text{ vapor}$$

Since the combustion occurs at constant volume, we need U_{rp}, which is derived with Eq. 1-7

$$U_{rp} = H_{rp} - 2480 \,(N_p - N_r)$$

N_p and N_r denote the product and reactant mole numbers for a chemically correct octane/oxygen mixture. Thus

$$U_{rp} = -5\,089\,100 - 2480(17 - 13.5)$$
$$= -5\,098\,000 \text{ kJ/kmol } C_8H_{18} \text{ vapor}$$

Now the mole numbers and the value for Q_v required in Eq. 5-30 can be set down as follows:

$Y \geqslant 12.5$	$8.5 \leqslant Y \leqslant 12.5$	
$N(1) = 0$	$N(1) = 2(12.5 - Y)CM$	
$N(2) = 8CM$	$N(2) = (2Y - 17)CM$	
$N(3) = 9CM$	$N(3) = 9CM$	5-33
$N(4) = 3.76Y \text{ CM}$	$N(4) = 3.76Y \text{ CM}$	
$(N5) = (Y - 12.5)CM$	$N(5) = 0$	

$$Q_v = -U_{rp}\frac{NM}{NMO} \qquad\qquad Q_v = -[U_{rp} + 2(12.5 - Y)281\,400]\frac{NM}{NMO}$$

where

$$CM = \frac{NM}{NMO} + \frac{NX}{NPO} \qquad\qquad 5\text{-}34$$

The value for Y fixes NMO and NPO by Eqs. 5-32; CM, which is simply a scale factor reducing the mole numbers to the proper size to fit in the engine, is known once NM and NX are known, or assumed in the case of NX, for the first cycle.

TABLE 5.2
IDEAL OTTO CYCLE; FULL THROTTLE CHEMICALLY CORRECT C_8H_{18}/AIR MIXTURE
$CR = 9$; VDISP = 0.823 liter; $P1 = 1$ atm; $TM = 300$ K.

Cycle	$T1$ (K)	$T2$ (K)	$T3$ (K)	$T4$ (K)	$T5$ (K)	$P2$ (atm)	$P3$ (atm)	$P4$ (atm)	KR	KP	$\dfrac{NM}{NM + NX}$
1	300	658	3167	1867	1295	19.8	101.0	6.6	1.36	1.24	1.00
2	328	720	3152	1857	1313	19.8	91.7	6.0	1.36	1.24	0.97
3	328	720	3153	1858	1313	19.8	91.7	6.0	1.36	1.24	0.97

Table 5.2 shows the rapid convergence of the computations. Corresponding power and thermal efficiency are in Table 5.3.

TABLE 5.3
IDEAL OTTO CYCLE; FULL THROTTLE*

Cycle	Power		η_{th}	MEP (atm)
	hp	kW		
1	47.4	35.6	45.1	17.0
2	42.6	31.8	45.1	15.3
3	42.6	31.8	45.1	15.3

*Same engine and operating conditions as in Table 5.2
(rpm = 3000).

By way of explanation, values of $T1$ were obtained with Eq. 5-29, while CPR was based on c_p values for the fuel vapor, air, and product gases at 300 K. Instead, of course, the variation of c_p for each gas with temperature may be incorporated into the calculations, using the values in Table E.1 for the coefficients a and b. The values for $NM/(NM + NX)$ show that after the first cycle, 97% of the reactant gases are fuel and air, with only 3% residual exhaust. It follows from Eqs. 5-20 and 5-27 that

$$\frac{NM}{NM + NX} = 1 - \frac{1}{CR}\frac{T1}{T5} \qquad \text{5-35}$$

The thermal efficiency, η_{th}, is calculated from

$$\eta_{th} = \frac{\text{WNET}}{-U_{rp}}\frac{NMO}{NM} \qquad \text{5-36}$$

The *mean effective pressure, MEP*, is defined as the pressure which, if acting throughout the power stroke, would produce the calculated work of that stroke, that is,

$$\text{WNET} = \text{MEP }\pi S\frac{B^2}{4}$$

so that

$$\text{MEP} = \frac{\text{WNET}}{\text{VDISP}} \qquad\qquad 5\text{-}37$$

Another way of examining the ratio of net work (or power) to displacement volume is to start with

$$\text{WNET} = \text{WEXP} - \text{WCOMP}$$

which may be expanded to

$$\text{WNET} = NP\{u_p(T3) - u_p(T4)\} - NR\{u_r(T2) - u_r(T1)\}$$

where NP and NR denote moles of products and reactants, respectively. But

$$NR = \frac{P1\ \text{VBDC}}{R\ T1} \qquad \text{and} \qquad \text{VBDC} = \frac{CR}{CR - 1}\ \text{VDISP}$$

hence

$$\frac{\text{WNET}}{\text{VDISP}} = \left[\frac{CR}{CR - 1}\frac{P1}{R\ T1}\right]\left[\frac{NP}{NR}\{u_p(T3) - u_p(T4)\} - \{u_r(T2) - u_r(T1)\}\right] \quad 5\text{-}38$$

The important feature in Eq. 5-38 is the fact that the compression ratio, pressure, temperature, and mole ratio, quantities that appear on the right-hand side, are intensive parameters; that is, the *size* of the engine does not influence them. This means, for example, that the statement in Table 5.2 that the calculations were obtained for an .823 liter displacement engine has no meaning, or rather, adds no meaning. All 9:1 compression ratio engines will exhibit the values in Table 5.2. By the same reasoning, all 9:1 compression ratio engines operating with chemically correct octane/air mixture and running full throttle at 3000 rpm, with 300 K intake manifold temperature, will deliver

$$\frac{42.6}{50.27} = 0.85\text{hp/in.}^3$$

or

$$\frac{31.8}{0.823} = 39.0\text{kW/liter}$$

But note carefully that Eq. 5-38 *does not* tell us how net work, or power when multiplied by one-half the rpm, changes as CR, $P1$, or $T1$ are changed. To examine these effects, cycle calculations must be carried through. Table 5.4 reveals the

TABLE 5.4
INFLUENCE OF COMPRESSION RATIO (CR) IDEAL OTTO CYCLE,
FULL THROTTLE CHEMICALLY CORRECT C_8H_{18}/AIR MIXTURE; 3000 rpm; VDISP = 0.823 liter; $P1$ = 1 atm;
TM = 300 K (RESULTS FROM THE THIRD CYCLE)

CR	T1 (K)	T3 (K)	T5 (K)	P3 (atm)	$\dfrac{NM}{NM+NX}$	Power		η_{th}
						hp	kW	
3	411	2944	1602	22.7	0.91	24.4	18.2	25.9
5	357	3042	1450	45.2	0.95	33.5	25.0	35.5
7	338	3105	1368	68.2	0.96	38.9	29.0	41.2
9	328	3153	1313	91.7	0.97	42.6	31.8	45.1
11	322	3194	1273	115.0	0.98	45.4	33.9	48.0

variation of engine performance with compression ratio. Compression ratio in a reciprocating piston engine is altered merely by relocating the position of the stroke. In rotary engines, compression ratio is changed by altering the contour of the rotor flank or by altering the volume of the depressions in the rotor or both.

As the compression ratio increases, the amount of residual exhaust gas in the engine during the compression stroke decreases, as does $T1$, the cylinder temperature at the start of compression. These two factors combine to produce an increase in power and thermal efficiency as the compression ratio increases. Compare the thermal efficiencies obtained by these calculations with the predictions of the air standard analysis, Eq. 5-8.

TABLE 5.5a
INFLUENCE OF FUEL/AIR RATIO IDEAL OTTO CYCLE,
FULL THROTTLE; 3000 rpm; VDISP = 0.823 liter; $P1$ = 1 atm; CR = 9; TM = 300 K

Y	$T1$ (K)	$T3$ (K)	$T5$ (K)	$P3$ (atm)	$\dfrac{NM}{NM + NX}$	Power hp	Power kW	MEP (atm)	η_{th}
9.5	327	2812	1128	88	0.97	39.7	29.6	14.3	32.1
10.5	327	2945	1199	90	0.97	40.9	30.5	14.7	36.5
11.5	328	3058	1260	91	0.97	41.8	31.2	15.0	40.8
12.5	328	3153	1313	92	0.97	42.6	31.8	15.3	45.1
13.5	328	3017	1255	88	0.97	39.8	29.7	14.3	45.5
14.5	327	2896	1204	84	0.97	37.4	27.9	13.4	45.9

Tables 5.5a and 5.5b illustrate the influences of the fuel/air mixture ratio on operating variables. In Table 5.5a, intake manifold temperature, TM, is constant, while in Table 5.5b, air temperature entering the carburetor, TA, is constant. Differences between the two tables reveal the effects of temperature drop resulting from fuel

TABLE 5.5b
INFLUENCE OF FUEL/AIR RATIO IDEAL OTTO CYCLE,
FULL THROTTLE; 3000 rpm; VDISP = 0.823 liter; $P1$ = 1 atm; CR = 9;
TA = 300 K

Y	TM (K)	$T1$ (K)	$P3$ (atm)	Power (hp)	η_{th}
9.5	272	297	96	44.0	32.2
10.5	274	300	97	44.8	36.5
11.5	276	302	97	45.5	40.9
12.5	278	305	98	46.0	45.1
13.5	280	306	93	42.8	45.6
14.5	281	307	88	40.0	45.9

evaporation, Eq. 5-7. Otherwise, engine conditions for the two tables are identical. Fuel/air ratio is indicated by Y, moles of oxygen/mole of C_8H_{18}, for which the chemically correct value is $Y = 12.5$.

Temperatures, pressures, and power peak at chemically correct mixtures, while thermal efficiency tends to creep upward for leaner mixtures.

Comparing these two tables at chemically correct mixtures, a change in $T1$, cylinder temperature at the start of compression, from 328 to 305 K increases power output from 42.6 to 46.0 hp. The effects of fuel evaporation are not insignificant. Low intake manifold temperatures can result in ice formation, which can cause havoc when chunks break off and damage engine parts, particularly supercharger compressor blades. This can be avoided by transferring heat from the exhaust manifold. Thus, TM, manifold temperature, is a controllable parameter.

The calculations and explanations up to this point apply to all Otto engines with carburetors or with fuel injection into the intake manifold. How will the analysis change if fuel is injected into the cylinder? Two changes are obvious: (1) TM, the manifold temperature, will equal ambient air temperature, TA and (2) CPR, the constant pressure heat capacity of the gases in the cylinder during compression, is based on air and residual exhaust gases. The energy equation will also change. Referring to Fig. 5.6, place a control volume around the cylinder, and the energy relation is

$$U_{\text{air+residual exhaust}}(T2) + h_{\text{fuel}}(T_{\text{fuel}}) = U_p(T3)$$

But

$$h(\text{liquid}) \approx u(\text{liquid})$$

and if the difference between $T2$ and T_{fuel} is ignored, the energy equation reduces to

$$U_r(T2) = U_p(T3) \qquad\qquad 5\text{-}39$$

which is identical to Eq. 5-17, except that the fuel is vapor in Eq. 5-17 and liquid in Eq. 5-39. The proper value for U_{rp} as it occurs in Eqs. 5-33 is

$$U_{rp} = H_{rp} - 2480\,(N_p - N_r)$$

as before, but now H_{rp} is the heat of reaction for liquid octane, and N_r is 12.5, not 13.5. Hence

$$U_{rp} = -5\,047\,800 - 2480\,(17\text{--}12.5)$$
$$= -5\,059\,000 \text{ kJ/kmol } C_8H_{18} \text{ liquid}$$

The change in U_{rp} is less than 1%, and the direct cylinder fuel injection calculations can be expected to show only small differences with the numbers in the tables given above.

5.6 IDEAL OTTO CYCLE: PART THROTTLE

The intake manifold pressure is reduced as the throttle closes, and the mass of fuel/air mixture drawn into the engine is reduced. The pressure-volume diagram now takes the form shown in Fig. 5.14, where PA denotes ambient atmospheric pressure and PM, intake manifold pressure. We need to analyze the exhaust and intake strokes, the path 4-5-6-1.

The temperature $T5$ is given, as before, by Eq. 5-19. The mass of residual exhaust that will remain in the engine for the next cycle is

$$NX = \frac{P5 \; VTDC}{R \; T5}, \qquad (P5 = PA) \qquad \qquad 5\text{-}40$$

and the amount of fresh charge at the start of the compression stroke is

$$NM = \frac{P1 \; VBDC}{R \; T1} - NX \qquad \qquad 5\text{-}41$$

and we need a value for $T1$.

Figure 5.15 illustrates three steps in the intake stroke. At TDC the fresh charge NM occupies volume VM in the intake pipe, and the pressure in the cylinder is $P5$. With the exhaust valve closed, the pressure in the cylinder will drop to $P1$ when the

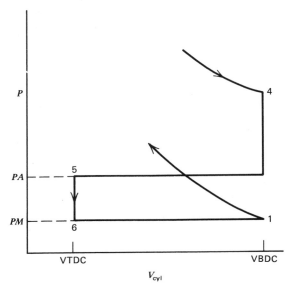

Figure 5.14 The exhaust/intake loop at part throttle.

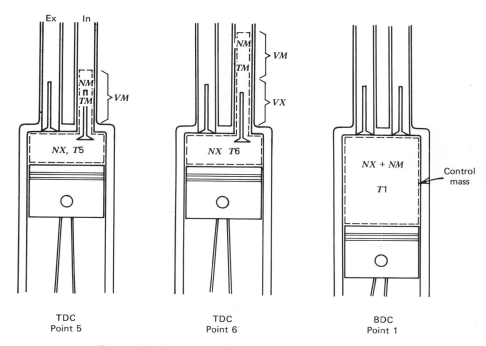

Figure 5.15 The exhaust/intake process at part throttle.

intake valve opens, reducing the temperature of the residual exhaust that remains in the cylinder to $T6$, given by

$$T6 = T5\left(\frac{P1}{P5}\right)^{(KP-1)/KP}$$ 5-42

Part of the exhaust will flow into the intake pipe, where it occupies volume VX, and will be cooler than $T6$, since work will have been done moving NM. The analysis is simplified when the temperature difference between the two portions of NX is ignored. The piston now moves slowly to BDC, drawing the escaped exhaust gas and NM into the cylinder. (The same analysis applies to rotary engines despite differences in geometry.)

An energy equation can be developed by repeating the steps used in Eqs. 5-21 through 5-24, and the result can be written in a form resembling Eq. 5-26, that is,

$$NX\,h_p(T6) + NM\,h_m(TM) = NX\,h_p(T1) + NM\,h_m(T1)$$ 5-43

Eliminating NX and NM with Eqs. 5-40 and 5-41, writing enthalpy terms as $c_p T$

products, and ignoring the difference between the exhaust and the mixture of fuel vapor, air, and exhaust (the latter cannot be evaluated until NX and NM are known), the following expression is obtained for $T1$:

$$T1 = \frac{CR\ TM}{CR + \dfrac{P5}{P1}\left[\dfrac{TM}{T5} - \dfrac{T6}{T5}\right]} \qquad\qquad 5\text{-}44$$

With a value for $T1$, NM may be assigned a value, and a new cycle may be carried through. The fraction of fuel/air mixture in the cylinder during the compression stroke will can be evaluated from Eqs. 5-40 and 5-41.

$$\frac{NM}{NM + NX} = 1 - \frac{1}{CR}\frac{P5}{P1}\frac{T1}{T5} \qquad\qquad 5\text{-}45$$

The fraction decreases as $P5$ remains fixed at ambient pressure, and $P1$ is reduced by closing the throttle.

One other item in the cycle calculation must now be changed; the work associated with the exhaust/intake loop enters as a term in evaluating the net work of the cycle. Defining,

WLOOP = work done on the gases in the cylinder by the piston during the exhaust and intake strokes,

$$= (P5 - P1)\ VDISP \qquad\qquad 5\text{-}46$$

and the expression

$$WNET = WEXP - WCOMP - WLOOP \qquad\qquad 5\text{-}47$$

should replace Eq. 5-18.

5.7 IDEAL OTTO CYCLE: SUPERCHARGED

Supercharging is achieved by installing a compressor in the intake system, thereby increasing the density of gases at the start of the compression stroke, which in turn increases the power output. Usually, the compressor is placed downstream from the carburetor or air-measuring device. This places less demand on the operation of either metering system. Supercharging may improve fuel vaporization by lengthening the path between the carburetor and the intake valve and by the temperature increase that accompanies the pressure rise in the compressor.

Supercharged cycles can be computed in a single pass through a computer program, for it is entirely reasonable to assume that with intake manifold pressure increased above ambient pressure in the exhaust manifold, all the exhaust gases are

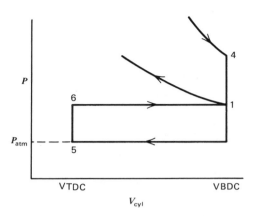

Figure 5.16 The exhaust/intake loop for a gear-driven supercharger.

swept out of the working space during the period of valve overlap, that is, the time interval when both valves are open or both ports are partially uncovered. However, if the period of overlap is excessive, the fuel/air mixture may be swept into the exhaust system. This has no effect on power output, but it will increase fuel consumption wastefully.

Power is required to drive the compressor and may be obtained by a direct mechanical drive from the crankshaft or by installing a turbine driven by the exhaust gases. Both cases will be examined.

Figure 5.16 illustrates the pressure-volume trace during the exhaust and intake strokes for the gear-driven compressor shown in Fig. 5.17. Following the notation in the latter figure, if TM denotes the temperature of the fuel/air mixture or air downstream from the carburetor or air-measuring system, the exit temperature from the compressor, $T1$, will be given by

$$T1 = TM \left[\frac{1}{\eta_c} \left\{ \left(\frac{P1}{PA} \right)^{(KR-1)/KR} - 1 \right\} + 1 \right] \qquad 5\text{-}48$$

where η_c is the isentropic compressor efficiency. The work required to drive the compressor, WSPCH, is then

$$\text{WSPCH} = NM \, CPR \, (T1 - TM) \qquad 5\text{-}49$$

NM denoting the moles of fuel/air mixture or air compressed, that is

$$NM = \frac{P1 \, \text{VBDC}}{R \, T1}$$

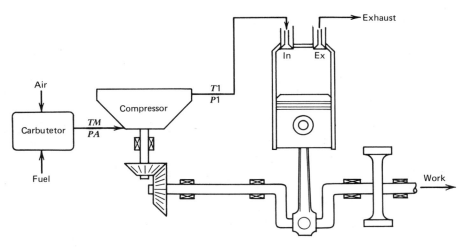

Figure 5.17 A gear-driven supercharger.

In calculating the net work of the cycle, WSPCH, must be included, as well as WLOOP, the work associated with the exhaust/intake loop (Fig. 5.16). If we define WLOOP as WLOOP = work done by the piston or rotor on the gases during the exhaust and intake loop

$$= (P1 - P5) \text{ VDISP} \qquad \qquad 5\text{-}50$$

then

$$\text{WNET} = \text{WEXP} - \text{WCOMP} - \text{WSPCH} + \text{WLOOP} \qquad 5\text{-}51$$

Table 5.6 lists some program output for a gear-driven supercharger. The power increase is 32% for a 50% increase in manifold pressure. The power increase depends, of course, on compressor efficiency.

TABLE 5.6
GEAR-DRIVEN SUPERCHARGER
CHEMICALLY CORRECT C_8H_{18}/AIR MIXTURE; CR = 9; VDISP = 0.823 liter;
TM = 300 K; 3000 rpm; η_c = 90%

$P1$ (atm)	$P3$ (atm)	$T1$ (K)	$T3$ (K)	$T5$ (K)	Power hp	kW	η_{th}	MEP (atm)
1.0	101	300	3167	1296	47.8	35.7	45.1	17.2
1.1	108	308	3181	1284	51.1	38.1	45.1	18.4
1.2	116	316	3194	1273	54.3	40.5	45.0	19.5
1.3	123	323	3206	1263	57.5	42.9	45.0	20.6
1.4	130	330	3218	1254	60.5	45.1	45.0	21.7
1.5	137	337	3229	1246	63.5	47.4	44.9	22.8

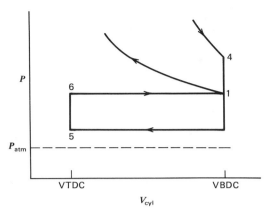

Figure 5.18 The exhaust/intake loop for an exhaust turbine supercharger.

Figure 5.18 is a sketch of the exhaust and intake strokes for a supercharger powered by an exhaust gas turbine, illustrated in Fig. 5.19. In Fig. 5.18 the exhaust stroke takes place at a pressure greater than ambient; otherwise there would be no driving force to move the turbine wheel. Figure 5.19 includes an exhaust control valve upstream from the turbine. The position of the valve determines what fraction of the

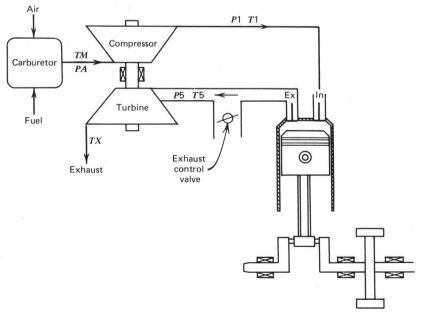

Figure 5.19 An exhaust turbine supercharger.

exhaust flow will pass through the turbine. The power output of the turbine is therefore controllable, a feature not available in the gear-driven supercharger illustrated in Fig. 5.17. To control the power input to the compressor, a gear box would be required, adding more machinery to the gear-driven system.

In the exhaust turbine system, the work required to drive the compressor is, as before,

$$NM \ CPR \ (T1 - TM)$$

while the work output of the turbine, assuming that all the exhaust passes through the turbine, is

$$w_{\text{turb}} = NP \ CPP \ (T5 - TX) \ \eta_t \qquad\qquad 5\text{-}52$$

TX denotes, as shown in Fig. 5.19, the exit temperature from the turbine for isentropic expansion, and η_t is the turbine efficiency. With PA denoting ambient pressure, the turbine work output can be brought into the form

$$w_{\text{turb}} = NP \ CPP \ T4 \left(\frac{PA}{P4}\right)^{(KP-1)/KP} \left[\left(\frac{P5}{PA}\right)^{(KP-1)/KP} - 1 \right] \eta_t$$

The required exhaust manifold pressure, $P5$, can be computed by equating compressor work input with the turbine work output,

$$\left(\frac{P5}{PA}\right)^{(KP-1)/KP} = 1 + \frac{NM \ CPR \ (T1 - TM)}{NP \ CPP \ T4 \ \eta_t} \left(\frac{P4}{PA}\right)^{(KP-1)/KP} \qquad 5\text{-}53$$

For an exhaust turbine system, net work is calculated from

$$WNET = WEXP - WCOMP + WLOOP$$

where

$$WLOOP = \text{work done by the gases on the piston}$$
$$= (P1 - P5) \ VDISP$$

as shown in Fig. 5.18.

The performance of an exhaust turbine supercharger engine, with all the exhaust products flowing through the turbine, is shown in Table 5.7. The gains in power output are very nearly the same for the two systems when all the exhaust flows through the turbine.

Table 5.7, however, is not a true example for an exhaust turbine system. Because it has been calculated on the assumption of full exhaust flow through the turbine, there is the tacit assumption that the resistance the turbine offers to gas flow can be

TABLE 5.7
EXHAUST TURBINE SUPERCHARGER
CHEMICALLY CORRECT C_8H_{18}/AIR MIXTURE; 3000 rpm; VDISP = 0.823 liter; CR = 9;
TM = 300 K; $\eta_{compressor} = \eta_{turbine}$ = 0.9

P1 (atm)	P4 (atm)	P5 (atm)	T3 (K)	T5 (K)	Power (hp)	η_{th}	MEP (atm)
1.0	6.6	1.00	3167	1296	47.8	45.1	17.2
1.1	7.1	1.03	3181	1284	51.3	45.2	18.4
1.2	7.6	1.06	3194	1273	54.7	45.4	19.6
1.3	8.0	1.08	3206	1263	58.1	45.5	20.9
1.4	8.5	1.11	3218	1254	61.4	45.6	22.0
1.5	9.0	1.13	3229	1246	64.6	45.7	23.2

altered by some means, thereby altering pressure $P5$ upstream from the turbine. In theory, this could be accomplished by changing the blade setting; in practice, this would be difficult to do and maintain the sort of reliability required. Practice follows the arrangement sketched in Fig. 5.19, where a control valve determines the quantity of exhaust passing through the turbine, with the remainder diverted directly to the atmosphere. When only a fraction f of the exhaust passes through the turbine, the equation for turbine work output is now, in place of Eq. 5-52,

$$w_{turb} = f\,NP\,CPP\,(T5 - TX)\eta_t \qquad 5\text{-}54$$

Then, in Eq. 5-53 for $P5$, f will appear in the denominator of the second part of the expression, with a resulting increase in $P5$. The net effect on power is to decrease WLOOP (see Fig. 5.18), thereby decreasing the power output. The inclusion of the f factor in the calculations does not present any computational difficulty; rather, the problem is one of knowing how a real turbine operates in conjunction with an exhaust control valve which can be positioned as we please. The value for f is fixed in accordance with the relative flow resistances of turbine and control valve, and we have a problem in fluid mechanics dealing with divided flow.

Note that increasing $P5$, which will occur with decreasing f values, not only reduces WLOOP, but when carried to the point where $P5$ exceeds $P1$, intake manifold pressure, the engine will no longer scavenge residual exhaust gases completely. One of the desirable features of supercharged engines is then lost. All of this means that turbines must be matched to engines, and to do the problem properly, power calculations require detailed knowledge of turbine characteristics as well as exhaust control valve geometry.

5.8 MODIFICATIONS TO THE IDEAL OTTO CYCLE

Figure 5.20 compares measured and calculated power output from a single-cylinder reciprocating piston engine. Measured power peaks at 3600 rpm. From Table

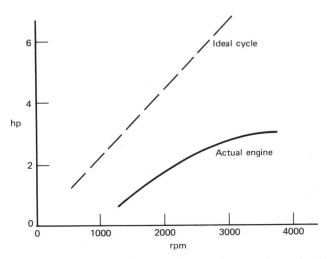

Figure 5.20 Comparison between measured and calculated power output. Single cylinder reciprocating engine. CR = 6.2; VDISP = 9.02 in^3.

5.4, at 6.2 compression ratio, power output is 0.74 hp/in^3, so calculated power at 3000 rpm is 6.7 hp, and the ideal cycle curve, a straight line passing through the origin, can be drawn.

Our interest is not with the relative placement of the two curves; calculated power is expected to exceed measured performance. The shape of the measured power curve is the feature of interest. All naturally aspirated (i.e., unsupercharged) engines, both rotary and reciprocating piston, exhibit a power peak at high speed. So, clearly, the ideal cycle analysis must be modified in some way to achieve closer approximation to the behavior of real engines. When the approach is confined to the methods and tactics of thermodynamics, three modifications are suggested: (1) the combustion process, (2) the intake and exhaust processes, and (3) heat transfer.

Combustion in spark ignition engines is not instantaneous. The spark is passed well before TDC, and combustion is not completed until well after TDC. The time required for combustion depends on a number of factors, the prominent ones being engine speed, spark plug location, combustion chamber size and geometry, and composition of the reactant gases. Thermodynamics suggests ways to deal with combustion that proceeds at some finite rate, although thermodynamics sheds no light whatsoever on the central question of the rate itself. Progressive combustion, as it is usually termed, is discussed in Section 5.9. Combustion chamber design is treated briefly in Section 5.10. (Combustion in Diesel-type engines proceeds rather differently from spark ignition combustion and is treated in Section 5.15.)

During the exhaust and intake processes the piston or rotor moves at finite speed and not, as we have assumed so far, infinitely slowly. Valves and ports open and close gradually, not instantly. The rates at which gases flow into and out of the working volume are controlled by pressure difference across valve and port, by the effective area of these openings, and by piston or rotor speed. Thermodynamics and fluid mechanics combine to provide adequate tools with which to make a relatively complete analysis possible. The processes described in Section 5.11 are discussed at a rather elementary level.

Heat transfer is present in all internal combustion engines, regardless of design or cycle. Power is reduced by the transfer of energy from the engine to the surroundings. It is a simple matter to formulate an expression for the rate of heat transfer; the problem, which is beyond the reach of thermodynamics, is to assign realistic values to the heat transfer coefficient which plays such a central role. Section 5.16 contains a brief formulation and discussion.

5.9 PROGRESSIVE COMBUSTION

Ideal Otto cycle calculations include the assumption of instantaneous combustion. In real engines, combustion occupies a finite time interval. Accordingly, as shown in Fig. 5.21, ignition occurs before the piston reaches top dead center, or the

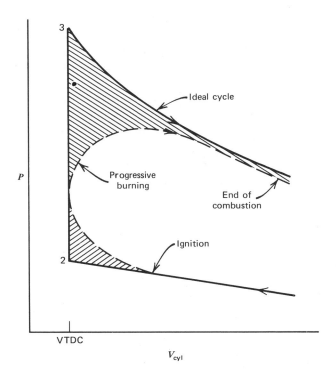

Figure 5.21

rotor reaches minimum volume, and combustion is not completed until the piston has moved beyond top dead center, or the rotor beyond the position for minimum volume. It is at once evident that the work done on the gases during the compression period will be increased and that work done by the gases during the expansion process will be decreased. Consequently, the effect of a finite time interval for combustion is to decrease power output. The objective of this section is to discover how to compute power output when the time required for combustion is included in the analysis. As we shall see, a quantitative treatment of some aspects of the combustion process is beyond the reach of thermodynamics. Therefore, we are forced to search for empirical results and experimental information.

In order to calculate the net work done by the gases between ignition and the end of combustion, we need to define where those two events occur in the cycle, and we need, in addition, the pressure-volume trace during combustion. The essential idea that underlies the anlysis of progressive combustion is expressed in the equation

$$\Delta P = \Delta P_r + \Delta P_c \qquad \qquad 5\text{-}55$$

where ΔP = pressure change during a small interval in time,

ΔP_r = pressure change during that time interval as a result of piston or rotor movement,

ΔP_c = pressure change during that time interval as a result of combustion.

Equation 5-55 merely asserts that these two separate contributions to the total pressure change are additive. Since compression, combustion, and expansion are assumed to proceed without heat transfer to or from the gases in the working space,

$$\Delta P_r = -\,PK\frac{\Delta V}{V}, \qquad K = \frac{c_p}{c_v} \qquad \qquad 5\text{-}56$$

In order to deal with the second term in Eq. 5-55, we must examine events that accompany combustion in a closed vessel. Suppose a combustible mixture is contained in a closed, rigid, and insulated vessel under known conditions of pressure and temperature. The mixture is ignited at some point. A burning front is immediately established and subsequently moves outward from the point of ignition in all directions. Eventually, all of the mixture is in the burned state at a pressure and a temperature that can be calculated by methods discussed in Chapter 2. If we define n as the mass fraction in the burned state,

$$n = \frac{\text{mass of burned gas}}{\text{total mass of gas}} \qquad \qquad 5\text{-}57$$

we want to answer the question, How does the pressure at any time in the combustion process vary with n? Pressure varies from initial to final value, while n passes from zero to unity. How are P and n related?

In the engine, piston motion accompanies combustion, so the working volume varies during the burning process. We first examine the relation betwen P and n for the case of constant volume combustion, after which the result can be very simply modified to include piston motion.

Referring to Fig. 5.22, let

P_i = initial pressure before the spark passes
T_i = initial temperature before the spark passes
P_e = pressure at the instant combustion ends
T_e = temperature at the instant combustion ends

We might note first that a temperature gradient will be established in the burned gas if there is no heat transfer between successively burning layers. The existence of that temperature variation can be seen by observing the behavior of the first and last piece to burn. Suppose a small parcel of reactant gas located between the spark electrodes is allowed to burn. Its temperature will be raised from T_i, before ignition, to $T_i + \Delta T_c$, the latter term denoting the temperature rise resulting from combustion. Since the parcel is small, it will burn at constant pressure. Consequently, the final temperature of the first piece to burn can be set down as

$$(T_i + \Delta T_c)\left(\frac{P_e}{P_i}\right)^{(k_p - 1)/k_p} \qquad\qquad 5\text{-}58a$$

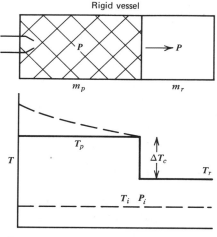

Figure 5.22 Progressive combustion in a rigid vessel.

where k_p denotes the specific heat ratio of the product gases. In contrast, the last piece to burn is first compressed isentropically from P_i to P_e, after which it burns, reaching a final temperature given by

$$T_i\left(\frac{P_e}{P_i}\right)^{(k_r-1)/k_r} + \Delta T_c \qquad\qquad 5\text{-}58b$$

The final temperature of the first piece to burn will be higher than that of the last piece. But we shall not retain the feature of a temperature gradient in the burned gas in the analysis that follows. For one thing, it complicates the algebra. More importantly, it seems entirely proper to assume that at some point in the expansion process, following the completion of combustion, the temperature gradient will be washed out by the gross motion of the gases moving about in the cylinder. We choose, therefore, to ignore the gradient throughout the burning process. (It turns out that when the gradient is retained, we also have to deal with the temperature rise resulting from combustion, ΔT_c, which is not, as implied in Eqs. 5-58a and 5-58b, constant; careful analysis reveals that ΔT_c changes in value as the combustion proceeds as a result of the difference between k_r and k_p).

Referring again to Fig. 5.22, at some arbitrary time in the combustion process, let P denote pressure, along with T_r and T_p, the temperatures of reactants and products, respectively. Denote by M the total mass of gas present and by m_r and m_p the masses of reactants and products. Since the entire combustion process is adiabatic, the energy relation between the initial state of affairs and some later time after the spark has passed is

$$Mu_r(T_i) = m_r u_r(T_r) + m_p u_p(T_p)$$

If we arbitrarily set

$$u_r(T_i) = 0$$

which is entirely legitimate, it follows that

$$u_p(T_e) = 0$$

and from the gas equation

$$T_e = \frac{P_e N_r}{P_i N_p} T_i \qquad\qquad 5\text{-}59$$

where N_r and N_p are the initial moles of reactants before combustion and the moles of products after combustion, respectively. The energy equation is now reduced to

$$m_r c_{vr}(T_r - T_i) + m_p c_{vp}(T_p - T_e) = 0$$

Let

$$a = \frac{c_{vr}}{c_{vp}}, \qquad b = \frac{N_r}{N_p}, \qquad \text{and} \qquad n = \frac{m_p}{M} \qquad\qquad 5\text{-}60$$

where n is the mass fraction in the burned state, as defined originally in Eq. 5-57. With Eqs. 5-60, the energy equation reduces now to

$$a(1 - n)(T_r - T_i) + n(T_p - T_e) = 0 \qquad\qquad 5\text{-}61$$

In addition to conservation of energy, we have specified that the volume of the vessel remains fixed during the combustion process, so that

$$\frac{MR_r T_i}{P_i} = \frac{m_r T_r R_r}{P} + \frac{m_p T_p R_p}{P}$$

which can be reduced to

$$b\left(\frac{P}{P_i}\right) T_i = n(1 - n)T_r + nT_p \qquad\qquad 5\text{-}62$$

We can eliminate T_p between Eqs. 5-61 and 5-62, and we know the reactants are compressed isentropically, so that

$$\frac{T_r}{T_i} = \left(\frac{P}{P_i}\right)^{(k_r - 1)/k_r}$$

Consequently, we reach the result we have been after by eliminating T_p and T_r, the product and reactant temperatures that prevail during the combustion process. Solving for n

$$n = \frac{(a - b)\left(\dfrac{P}{P_i}\right)^{(k_r - 1)/k_r} + b\dfrac{P}{P_i} - a}{(a - b)\left(\dfrac{P}{P_i}\right)^{(k_r - 1)/k_r} + b\dfrac{P_e}{P_i} - a} \qquad\qquad 5\text{-}63a$$

For a typical combustion engine reactant mixture, $a \approx .7$ and $b \approx .95$ are reasonable values. Thus Eq. 5-63a can be reduced to something more manageable, namely,

$$n = \frac{bP - aP_i}{bP_e - aP_i} \qquad\qquad 5\text{-}63b$$

or an approximate and still simpler form can be written:

$$n = \frac{P - P_i}{P_e - P_i}$$

5-64

Note that Eq. 5-64 would be exact if there were no difference in thermodynamic properties between reactants and products and no mole number change resulting from combustion. In the discussions that follow we shall use the simple form, Eq. 5-64. The exact form, Eq. 6-63a, can be used readily, since a and b, Eq. 5-60, are both known. However, the differential of the exact equation is unwieldy. Since we need the differential dn, the computational method is presented with greater clarity when we use Eq. 5-64.

Before proceeding further, we might note the relation between the volume occupied by the product gases, V_p, and the volume of the vessel, V. Using Eq. 5-64, the ratio of the two volumes can be expressed in the form

$$\frac{V_p}{V} = 1 - \frac{1 - n}{[1 + n(P_e/P_i - 1)]^{1/k_r}}$$

5-65

By way of illustration, with $P_e/P_i = 4.5$ and $k_r = 1.35$

n	V_p/V
0.1	0.28
0.2	0.46
0.3	0.59
0.4	0.69
0.5	0.76
0.6	0.83

When half of the reactants have been converted to products, that is, $n = .5$, the products occupy 76% of the volume of the vessel, reflecting the large increase in volume that accompanies combustion.

Equation 5-65, or the counterpart obtained with Eq. 5-63a or 5-63b in place of Eq. 5-64, is useful in engine studies. Because the product gases are hot and luminous, a photographic history of the combustion process allows the mass fraction in the burned state, n, to be computed and displayed against crank angle or time.

Continuing with the analysis of burning, the term ΔP_c in Eq. 5-55 is, using Eq. 5-64,

$$\Delta P_c = (P_e - P_i) \, \Delta n$$

Recall now that P_i and P_e are the initial and final pressures for constant volume combustion. If, as in Fig. 5.21, combustion had occurred at TDC, then $P_i = P2$ and $P_e = P3$. Consequently, when combustion proceeds at some volume V, or when a fraction Δn of the reactants burn at volume V, the relation between ΔP_c and Δn is simply written as

$$\Delta P_c = (P3 - P2) \frac{\text{VTDC}}{V} \Delta n \qquad \text{5-66}$$

Note that $P2$ and $P3$ are calculated from the sharp-cornered *ideal* Otto cycle.

Combining Eqs. 5-55, 5-56, and 5-66, we arrive at the main result:

$$\boxed{\Delta P = - PK \frac{\Delta V}{V} + (P3 - P2) \frac{\text{VTDC}}{V} \Delta n} \qquad \text{5-67}$$

Equation 5-67 relates the change in cylinder pressure ΔP to the volume change ΔV and the mass fraction of gas that burned during the time interval required for ΔV. Since the difference between KR and KP, the ratio of the specific heats of reactants and products, respectively, is not insignificant, the integration of Eq. 5-67 can be carried out with the relation

$$K = KR + (KP - KR)\, n \qquad \text{5-68}$$

There is no logic for Eq. 5-68; it merely provides a simple and convenient transition from reactants to products as the combustion process proceeds.

The task we now face is integration of Eq. 5-67, which provides pressure, P, and volume, V, values as combustion proceeds from ignition to completion. With those numbers in hand, we can compute the net work done by the working gases during the combustion period. We define

$$\theta_i = \text{crank angle (or rotor angle) at ignition}$$
$$\Delta\theta_c = \text{span of crank angle (or rotor angle) for the combustion period}$$
$$\theta_e = \text{crank angle (or rotor angle) when combustion ends}$$
$$= \theta_i + \Delta\theta_c$$

One way to carry out the integration is to divide the combustion period into N steps, each occupying uniform intervals of time or angular motion. Thus

$$\Delta\theta = \frac{\Delta\theta_c}{N}$$

Given the geometry of the piston-crank or rotor-bore combination, the volume change ΔV in Eq. 5-67 is simply

$$\Delta V = V(\theta + \Delta\theta) + V(\theta) \qquad\qquad 5\text{-}69$$

where $V(\theta)$ is a known algebraic expression, Eqs. 5-4 and 5-6, for example. But now we need to know how n, the mass fraction in the burned state, varies with θ, and here thermodynamics is of no help to us. That science deals with equilibrium states, whereas the dependence of n on θ involves the succession of a series of equilibrium states that change with time. This leaves us with two alternatives: we can think about the combustion process and propose hypothetical relations between n and θ which seem to us reasonable and possible, or we can examine as best we can the events that take place in real engines, hoping to be led in this way to empirical relations. This latter method, the experimental approach, is appealing but unfortunately meets with great difficulties when pursued actively in the laboratory. Combustion is a rapid chemical reaction of enormous complexity, and in engines it takes place in a relatively small space that is not easily accessible to measuring apparatus.

What happens following passage of the spark in a combustion engine? Photographic studies taken through quartz cylinder heads reveal what we would expect; a flame front develops rapidly and moves away from the spark plug, or spark plugs, in radial directions. The rate at which that flame front converts reactants to products depends, among other things, on two obvious factors: the momentary surface area of the flame front, and the speed with which that front moves relative to the adjacent unburned gas. Clearly, the momentary area of the flame front depends on the geometry of the combustion chamber and the location of the spark plug, or locations of the plugs if more than one is provided. We could assume a relationship of the following form:

$$n = \frac{\theta - \theta_i}{\Delta\theta_c}, \qquad \text{uniform rate}$$

so that   5-70

$$\Delta n = \frac{\Delta\theta}{\Delta\theta_c} = \frac{1}{N}$$

Equation 5-70 recommends itself, if only for its simplicity. Whether it bears any relationship to real events in a real engine is largely a matter of geometry. Equation 5-70 might be a good approximation for a rotary engine. On the other hand, it might be a poor rule to apply to a piston engine. For either engine, it is not realistic during the early stage of combustion, for the flame front starts as a tiny sphere between the spark electrodes, and therefore $\Delta n/\Delta\theta$ should be very small for $\theta \approx \theta_i$. Equation 5-70 does not fit that requirement.

A second possibility can be written as

$$n = \left[\frac{\theta - \theta_i}{\Delta\theta_c} \right]^2, \qquad \text{square law}$$

so that   5-71

$$\Delta n = 2 \frac{\theta - \theta_i}{\Delta\theta_c} \frac{\Delta\theta}{\Delta\theta_c} = \frac{2}{N} \frac{(\theta - \theta_i)}{\Delta\theta_c}$$

That equation satisfies the condition of slow initial growth of the flame front, but it predicts continuous growth of the flame front. Neither a rotary engine nor a piston engine permits continuous flame front area increase. In a piston engine, the advancing flame front will in time reach the cylinder head and the piston face, either of which will cause an abrupt decrease in the rate of flame front area change with time.

The third possible relation can be written

$$n = \frac{1}{2} \left[1 - \cos \frac{\theta - \theta_i}{\Delta\theta_c} \pi \right], \qquad \text{empirical}$$

and   5-72

$$\Delta n = \frac{\pi}{2N} \sin \frac{\theta - \theta_i}{\Delta\theta_c} \pi$$

This relationship has been derived from engine studies. Note that $\Delta n/\Delta\theta$ is small at the end, as well as at the beginning of combustion.

Figure 5-23 provides a graphic comparison of these three relationships. Which one should we choose? One way to answer this question is to ask another: To what extent is power influenced by the choice among the three relationships? We shall examine this question momentarily.

The relationship between n and θ is only a part of our conjecture about the combustion process. Each relationship contains a term $\Delta\theta_c$, the angular span of crank of rotor movement between ignition and the end of combustion. What value should we assign to it? For an answer, we must turn to experimental results. Early experimentation with photographic studies yielded numbers for $\Delta\theta_c$, although that method lacked precision. The onset and termination of combustion are not, as one might suppose, easy to identify from photographs. Present practice is to install a set of sensors in the cylinder head, using sensors that respond in some simple way to the passage of the flame front. Later in this section, and also in Section 5.10, we examine $\Delta\theta_c$ on a quantitative basis. For the moment, we can pose a second question, which follows the question raised above: To what extent is power influenced by the value assigned to $\Delta\theta_c$? The answers to these two questions, taken separately and together, will provide clues with which to judge the importance of the rate of combustion (n as

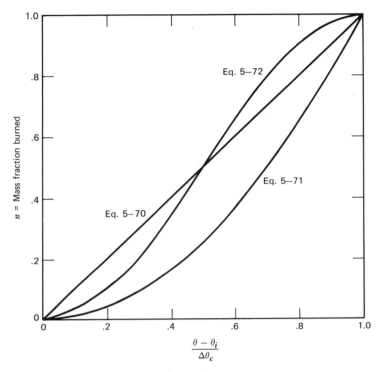

Figure 5.23

a function of θ) and the duration of combustion ($\Delta\theta_c$) on power output and other engine parameters.

Cycle calculations for an engine operating with progressive-type combustion can be carried out in the following manner: Start at bottom dead center (the description given here will be couched in piston engine terms) with assumed values for temperature $T1$ and residual exhaust gas mole number NX prior to compression. With NX, the specific heat ratio of the reactant mixture, KR, can be obtained. Now carry the *ideal cycle* through to the end of constant volume adiabatic combustion, obtaining the values for $P2$ and $P3$ (see Fig. 5.21), which are required in Eq. 5-67. With $T3$, the ratio of the specific heats for the product gases, KP, can be obtained.

Choose a spark advance number, SA, in degrees before top dead center. Then with subscript i denoting values at ignition,

$$\theta_i = 180 - SA$$
$$V_i = V(\theta_i)$$

$$P_i = P1\left(\frac{\text{VBDC}}{V_i}\right)^{KR}$$

$$T_i = T1\left(\frac{\text{VBDC}}{V_i}\right)^{KR-1}$$

and the work required for the compression stroke is now

$$\text{WCOMP} = (NM + NX)\, CVR(T_i - T1) - \Sigma\left(P + \frac{\Delta P}{2}\right)\Delta V \qquad 5\text{-}73$$

where CVR represents the constant volume specific heat of the reactant mixture $NX + NM$. The summation is carried out from ignition to top dead center, using equal increments in crank angle with which the corresponding increments in V, P, and n may be calculated, using Eqs. 5-67 and 5-68.

Continuing beyond top dead center, with subscript e identifying the end of combustion, P_e and V_e are known, since the end of combustion is fixed by the assumed value for $\Delta\theta_c$. Then

$$T_e = \frac{P_e V_e}{R\, NP}$$

$$T4 = T_e\left(\frac{V_e}{\text{VBDC}}\right)^{KP-1}$$

where NP represents the moles of product gas, and $T4$ represents the temperature at bottom dead center, at the completion of the expansion stroke. The work of expansion is

$$\text{WEXP} = \Sigma\left(P + \frac{\Delta P}{2}\right)\Delta V + NP[u_p(T_e) - u_p(T4)] \qquad 5\text{-}74$$

where the summation extends from top dead center to the end of combustion.

Equation 5-67 is integrated numerically in stepwise fashion. The accuracy of the calculations depends on the number of steps, N, selected. Consequently, several values for N should be tried in order to acquire some feel for the size that will be adequate but not excessive.

The balance of the cycle, the exhaust and intake strokes, is completed by the methods of Section 5.6. Because the cycle calculation, and an engine, start with assumed values for some parameters that differ from the values at steady operation, the cycle must be completed more than once. The computations are halted when successive cycles repeat one another.

We can now examine the influences of rate of combustion and duration of combustion as they are revealed by computations. Figure 5.24a illustrates the sort of results obtained when combustion duration and spark advance are varied independently. In this example, the rate of combustion was chosen as the linear model, Eq. 5-70. The figure shows that power is influenced markedly by combustion duration and spark advance. For a given combustion duration, $\Delta\theta_c$, best power is obtained when combustion divides about evenly between the periods before and after top dead center position. Real engines exhibit the same feature, or very nearly so. As combustion duration increases, power decreases. This engine, operating on the ideal cycle, produces 42.6 hp (Table 5.3). With $\Delta\theta_c = 90°$, for example, maximum available power is reduced to 34.5 hp.

In Fig. 5.24b, combustion duration is fixed at $60°$, and the curves show how the three models for rate of combustion differ from one another in computed power output. The ordinate in that graph has been stretched out to accentuate differences.

Figure 5.25 is a plot of P against θ for the engine of Fig. 5.24a, with $30°$ spark advance and $60°$ combustion duration, operating with three different rates of combustion. Note the differences in peak cylinder pressure between these three burning rates. Curves of the type shown in Fig. 5.25 can be obtained with modern instrumentation from engines, and working backward, through Eq. 5-67, the relationship between n and θ can be plotted and worked into an algebraic approximation.

Up to this point in the discussion of Otto cycle engines the net work output for any cycle has been assumed to be independent of engine speed. When the notion of progressive combustion is introduced, we have to include the duration of combustion, $\Delta\theta_c$, in the computations, since, as illustrated in Fig. 5.24a, for a fixed engine speed, power is strongly dependent on the value assigned to $\Delta\theta_c$. The question then arises, What is the dependence of $\Delta\theta_c$ on engine speed?

Guided by common sense alone, we conclude that $\Delta\theta_c$ is directly proportional to speed. That is, if combustion extends over $20°$ crankshaft rotation at, say, 1000 rpm, it will occupy $40°$ at 2000 rpm, $60°$ at 3000 rpm, and so on. This conclusion would be based on the notion that the *time interval* for combustion is constant, independent of engine speed, for a given fuel and a given fuel/air ratio. Indeed, the time interval would be constant if the gases in the combustion chamber were totally at rest the instant the spark is passed. In fact, the gases are not at rest; instead, they swirl about in patterns determined by combustion chamber geometry and piston speed. The effect of this motion seems to be to induce large-scale turbulence, which in turn exerts an effect, probably in some complicated way, on the rate of burning. For one thing, turbulence may enlarge the burning front area, which in turn will reduce the time required for combustion to turn its full course. At this point, and in this book, it is quite useless to tarry over the question. The theoretical analysis is quite beyond our scope, and we turn instead to the results of experiment, which are rather abundant, albeit clouded by some uncertainty as to precision. For one thing, turbulence in the region of the spark plug may delay the establishment of a burning front that can be

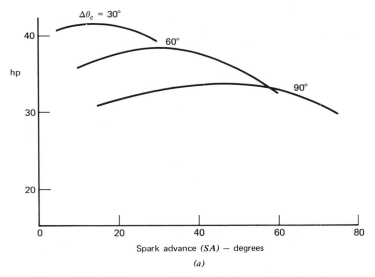

Figure 5.24a Influences of spark advance and combustion.

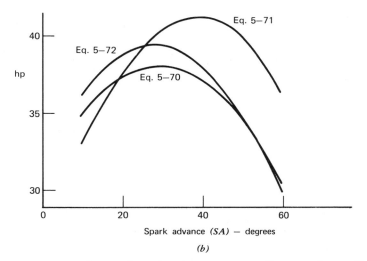

Figure 5.24b Influence of rate of combustion on power. (Same engine as Fig. 5.24a.)

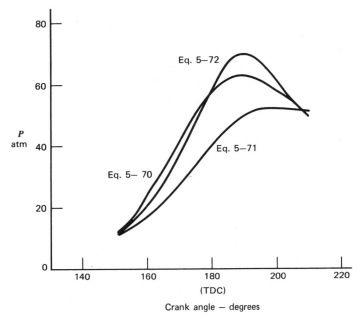

Figure 5.25 Same engine as Fig. 5.24a, 30° spark advance, 60° combustion duration.

sustained. Experiments reveal cycle-to-cycle variations in power output, which in some cases lead to misfiring when a burning front never is established.

For an automotive-size engine, the following expression, based on experiment,* is a reasonable representation of connection between $\Delta\theta_c$, speed (rpm) and in addition, fuel/air for a gasoline-fueled engine, stated as Y_{cc}/Y, with Y denoting moles of oxygen and Y_{cc} denoting chemically correct oxygen requirement:

$$\Delta\theta_c = 40 + 5\left(\frac{\text{rpm}}{600} - 1\right) + 166\left(\frac{Y_{cc}}{Y} - 1.1\right)^2 \text{(degrees)} \qquad 5\text{-}75$$

Applied to the engine described in Fig. 5.24a, the variation of power with speed is displayed in Fig. 5.26, where the three expressions for rate of combustion have been included and the ideal cycle (42.6 hp at 3000 rpm) shown for comparison. For all three rates of combustion, the departure of calculated power from the linear relation that characterizes all ideal cycle computations becomes more pronounced as speed increases. Figure 5.26 illustrates how calculated engine performance begins to imitate

*See C. F. Taylor, *The International Combustion Engine in Theory and Practice*, M.I.T. Press, Cambridge, Mass., 1977, Volume II, Chapter 1. This superb publication contains an enormous amount of experimental data on all aspects of engine performance.

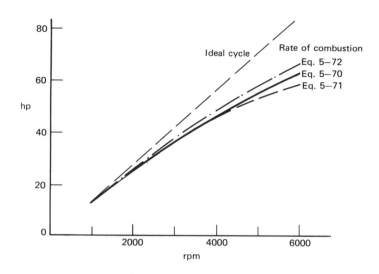

Figure 5.26 Power versus Speed. CR = 9, TM = 300 K, VDISP = .823 liter, P1 = 1 atm. Full throttle. C_8H_{18} fuel; Y_{cc} = 12.5, T = 10.5. $\Delta\theta_c$ from Eq. 5-75, spark advance fixed at 20° BTDC.

real engine performance as the progressive nature of the combustion process is brought into the analysis.

The agreement between theory and practice is improved further when piston speed and valve geometry are included in the analysis. Those features are discussed in Section 5.11. However, we shall first briefly examine the role of combustion chamber design in engine performance.

5.10 COMBUSTION CHAMBER DESIGN IN SPARK IGNITION ENGINES

The question we would like to answer is as follows: Given the shape and dimensions of a combustion chamber, spark plug location, and pressure, temperature, and mixture composition (fuel, air, residual exhaust), what will be the time interval between spark and the end of combustion? We cannot answer the question definitively. In this section, we undertake a brief look at the problem, indicating the part that thermodynamics plays in the solution.

Consider first Fig. 5.27a, a long tube, open at one end and filled with combustible mixture. When a spark is passed at the *open* end, a flame front will develop and move to the right. Since hot products are free to leave through the open end, pressure will remain essentially constant. If we assume little or no heat transfer from

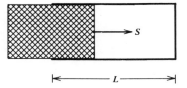

(a) A long tube, open at one end

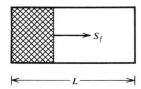

(b) A long tube, closed at both ends

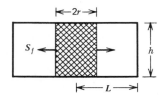

(c) A right cylinder, radius L

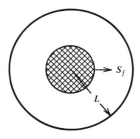

(d) A sphere, radius L

Figure 5.27 A variety of chambers with simple geometry.

the flame front to the unburned gas (the flame will radiate vigorously because of its high temperature, however), then the state of the unburned gas remains undisturbed, and we would expect the flame front to move at a constant speed, S, referred to as the *normal burning velocity*. The adjective "normal" enters the definition; to an observer moving with the flame front, unburned gas would appear everywhere to approach with S as the velocity component normal to the surface. We cannot predict S from first principles; we can measure it, however. Figure 5.27*a* could be used to make the measurement. If we know S, the combustion time is L/S.

Now consider Fig. 5.27*b*, the tube with both ends closed. The mixture is ignited at the left end, and a flame front moves to the right at speed S_f, the *flame speed*. Because of the expansion that accompanies combustion, the flame front is driven to the right by the hot products, and the unburned gas moves to the right also. Denoting the velocity of the reactants at the flame front by S_r,

$$S = S_f - S_r$$

from the definition for S. Thus the flame speed will be larger than the normal burning velocity; the two are equal at the end of combustion, since the last piece to burn is against the wall and cannot move, that is, $S_r = 0$.

For the right cylinder (Fig. 5.27*c*) and the sphere (Fig. 5.27*d*), a flame front will move with speed S_f, but for the three closed vessels, S_f may take on different values, even for the case of S constant. We can now ask, What are the burning times for the closed chambers? If we assume S constant, we can formulate the problems and answer the question.

Using the definition for S, the rate at which products will be formed is

$$\frac{dm_p}{dt} = \rho_r A S$$

where subscripts p and r denote products and reactants, ρ is the density, A is the flame surface area, and t is time. Introducing the total mass of gas, M, and eliminating the density, rearrangement leads to

$$S = \frac{MR_r T_i}{AP_i} \frac{P_i}{P} \frac{T_r}{T_i} \frac{dn}{dt}$$

Subscript i denotes ignition, and n is the mass fraction in the burned state. With no heat loss to the surroundings, the reactants are compressed isentropically,

$$\frac{P_i}{P} \frac{T_r}{T_i} = \left(\frac{P}{P_i}\right)^{-1/k_r}$$

We can eliminate P in favor of n by using the relation found in the previous section, Eq. 5-64,

$$\frac{P}{P_i} = n\left(\frac{P_e}{P_i} - 1\right) + 1$$

Making these substitutions,

$$S = \frac{MR_r T_i}{A P_i} f(n) \frac{dn}{dt}$$

or

$$S = f(n) \frac{V}{A} \frac{dn}{dt}$$

and

$$f(n) = \left[n\left(\frac{P_e}{P_i} - 1\right) + 1 \right]^{-1/k_r} \qquad 5\text{-}76$$

where V is the total volume of the combustion chamber. Equation 5-76 is the general equation and can be applied to any vessel. We now apply it to the three closed volumes in Fig. 5.27. Take the closed tube first, and assume that A is constant. Then

$$t = \frac{L}{S} \int_0^n f(n)\, dn \qquad 5\text{-}77$$

which can be integrated. We are for the moment interested in t_c, the combustion time, which is found by setting $n = 1$. The result is

$$t_c = \frac{L}{S} \frac{k_r}{(k_r - 1)(P_e/P_i - 1)} \left[\frac{P_e}{P_i}^{(k_r - 1)/k_r} - 1 \right] \qquad 5\text{-}78$$

which can be evaluated when k_r, the specific heat ratio of the reactants, and P_i and P_e, the initial and final pressures, are known.

Turning next to the right cylinder, Fig. 5.27c, the area of the burning front is

$$A = 2\pi r h$$

where r and h are indicated in the figure; we assume h is small compared to L, and we neglect the curvature of the burning front that would result from point-ignition. Since

$$V = \pi L^2 h$$

Eq. 5-76 reduces to

$$\frac{S}{L} = \frac{1}{2}\frac{L}{r} f(n) \frac{dn}{dt}$$

But from Eq. 5-65

$$\left(\frac{r}{L}\right)^2 = \frac{V_p}{V} = 1 - (1-n)f(n)$$

where V_p denotes the volume of the products of combustion. Then for the right cylinder

$$t_c = I\frac{L}{S}, \qquad I = \frac{1}{2}\int_0^1 \frac{f(n)}{\sqrt{1-(1-n)f(n)}} \, dn \qquad\qquad 5\text{-}79$$

Numerical integration is required.

For the sphere with central ignition, with r denoting the radius of the burning front,

$$A = 4\pi r^2 \qquad \text{and} \qquad V = \tfrac{4}{3}\pi L^3$$

so that Eq. 5-76 for this case reduces to

$$\frac{S}{L} = \frac{1}{3}\frac{L}{r} f(n) \frac{dn}{dt}$$

and because

$$\left(\frac{r}{L}\right)^3 = \frac{V_p}{V} = 1 - (1-n)f(n)$$

the case of the sphere reduces to

$$t_c = I\frac{L}{S}, \qquad I = \frac{1}{3}\int_0^1 \frac{f(n)}{[1-(1-n)f(n)]^{2/3}} \, dn \qquad\qquad 5\text{-}80$$

To make a numerical comparison, take $P_e/P_i = 4.6$, and $k_r = 1.35$, which are reasonable values, and then the combustion times compare as follows:

(a) Tube, open at one end $t_c = L/S$
(b) Tube, closed both ends $t_c = 0.52 \, L/S$ 5-81
(c) Cylinder, central ignition $t_c = 0.38 \, L/S$
(d) Sphere, central ignition $t_c = 0.29 \, L/S$

The purpose of this exercise has been to demonstrate the dependence of burning time on chamber geometry. With identical flame travel distance from ignition point to remotest point, different geometries produce different burning times because geometry controls the ratio of chamber volume, V, to burning surface, A.

The relations between mass fraction burned and t/t_c is shown in Fig. 5.28, where it is evident that the tube, the cylinder, and the sphere lie close to one another. Note the good approximation of the square law for the cylinder and sphere. Note also that the latter two curves begin with horizontal slope, because with point-ignition, A is initially zero. In the case of the tube, the entire cross section was assumed to be ignited, which is unrealistic and impossible with point-ignition.

For the same vessel, the location of the spark plug may be varied over considerable limits. Referring to Fig. 5.29a and b, with central ignition in a tube closed at both ends, the burning area is doubled, and the burning time is therefore halved, bringing it down to the value for a sphere. Tubes, however, are academic for piston/cylinder engines (but do apply to rotary engines, which we discuss shortly). With the cylinders,

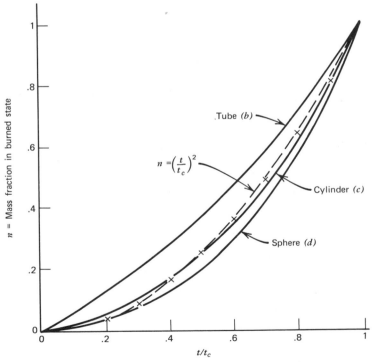

Figure 5.28 Relation between n and t/t_c, for the three closed vessels in Fig. 5.27.

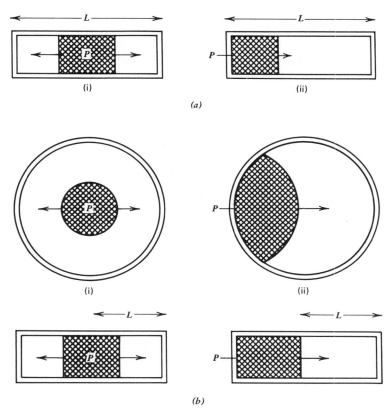

Figure 5.29 Two examples of simply shaped vessels with different spark plug locations (*P*). (*a*) Tubes with both ends closed: (i) central ignition; (ii) end ignition. (*b*) Cylinders: (i) central ignition; (ii) side ignition.

which do represent one form of reciprocating piston combustion chamber, a wide variety of spark plug locations are at the disposal of the designer. We consider only two, central ignition and side ignition (Fig. 5.29*b*). The case of central ignition has already been worked out in Fig. 5.27 and Eq. 5-79. For side ignition, the relations between flame area, burned gas volume, and total volume can be worked out readily and the results integrated numerically, with the following results:

$$\begin{array}{ll}
\text{Cylinder, central ignition} & t_c = .38 \, L/S \\
\text{Cylinder, side ignition} & t_c = 1.04 \, L/S
\end{array}$$

The burning times differ by a factor of 2.7. In this case, it is instructive to plot *n* versus *t*, rather than *n* versus t/t_c, for reasons that are apparent from Fig. 5.30. Both

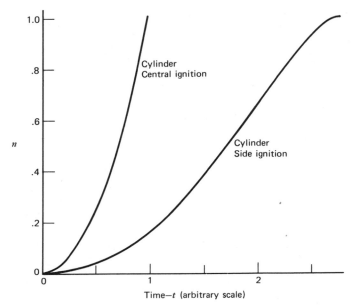

Figure 5.30 Mass fraction burned, **n**, as a function of time, for the two cylinders in Fig. 5.29*a* and **b**.

curves show horizontal slopes at $n = 0$ because of point-ignition. With central ignition, the area of the burning surface increases continuously. With side ignition, the burning surface area will reach a maximum and then decrease, going to zero as it approaches the far wall. During that close approach, there is a marked decrease in the rate of burning. Note, for example, the comparative times for the two systems between $n = 0.9$ and $n = 1$; the ratio of the two is about 7:1. This has large implications with regard to the tendency toward detonation, a topic discussed in Section 5.13.

The S-shape of the n versus t curve for the cylinder with side ignition is reminiscent of the n versus θ expression in Eq. 5-72, plotted in Fig. 5.23. That empirical expression is a very good fit for side ignition but not for central ignition.

Figure 5.31 illustrates the position of the rotor of a planetary-type rotary engine near minimum volume. The usual arrangement is to fire the spark plug when some part of the flank depression is opposite the plug. The depression then provides a connecting passage between the two parts of the combustion chamber that are formed by the necking-in of the bore along the minor axis. In rotary engines the designer clearly has a wider choice of spark plug locations than in reciprocating engines. In addition, the contour of the depression in the flank can be shaped in some purposeful manner to assist flame movement. Note also that the rotor moves relative to the bore during combustion, which is likely to induce more gas movement than will be found in

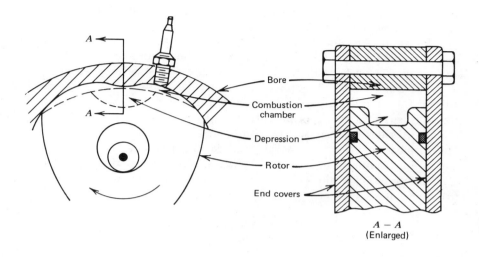

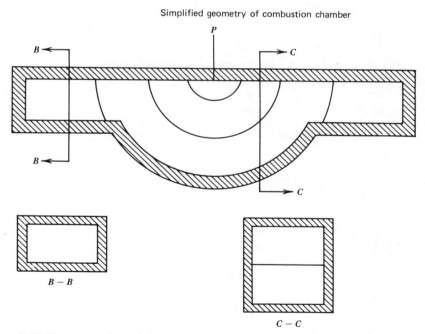

Figure 5.31 Rotary engine, planetary rotation type, showing rotor position near point of ignition, and a simplified geometric model of the combustion chamber.

reciprocating piston combustion chambers. Equation 5-76 applies to rotary engines in exactly the same way as it was applied to the combustion chambers in Fig. 5.27.

Equation 5-76 could be used to examine combustion progress in a rotary engine, using the simplified model in Fig. 5.31, and it could be used to study the difference between spark plug locations on combustion in a dome-head engine, shown in Fig. 5.32. For either engine, the procedure assumes that the flame front moves radially outward from the spark plug into the combustible gas. At any moment, the flame surface is part of a sphere, or portions of a sphere, with its center at the spark plug location. A series of flame surfaces can be drawn and their areas, A, determined. If total chamber volume, V, is known, then the ratio V/A which occurs in Eq. 5-76 is known. For each flame surface, the volume that it encloses, V_p, is evaluated, and with V_p/V known, we can, with Eq. 5-65, find the value of n and thus fix the number for $f(n)$. Now plot

$$f(n)\frac{V}{A} \text{ against } n$$

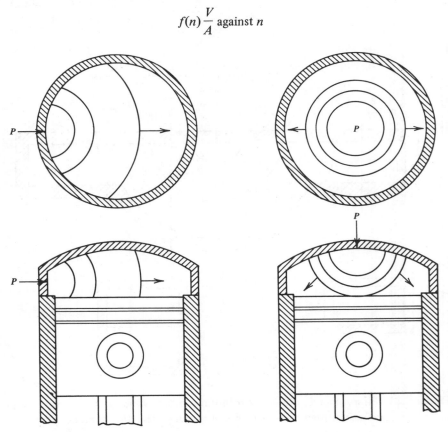

Figure 5.32 Dome-head combustion chamber, with two spark plug locations.

and by numerical integration, the combustion time, t_c, is found in terms of S, the normal burning velocity.

This exercise, while straightforward, is likely to be time-consuming and is certainly tedious, so one may well ask, What can we learn from all that labor? To answer the question, we recall the assumptions that underlie Eq. 5-76.

1. Constant volume,
2. S, normal burning velocity, constant,
3. Combustible mixture quiescent at ignition.

With regard to assumption 1, the piston or rotor moves during the time required for combustion, but because of engine geometry, the associated volume change is rather small. Assumption 1 is therefore reasonable. As for assumption 2, the normal flame velocity is known to vary with fuel, reactant composition, pressure, and temperature. However, our understanding of the dependence of S on these parameters is incomplete. The variations with temperature and reactant composition are particularly strong. Assumption 2 is wide open to question.

With regard to assumption 3, the assumption of a quiescent mixture at the instant of ignition is totally unrealistic. The reactants may reach high velocity as they enter the working space, and the velocity more or less persists during the compression stroke. At ignition there is considerable movement in the gases in the engine. From measurements it is well known that the normal burning velocity is much too small to account for the rapid rate of combustion observed in engines fitted with windows for viewing the progress of the flame front. The best explanation for short burning time is the role played by turbulence. The flame surface is not a smooth spherical section; instead, turbulence produces a more complicated flame surface with a total area *much larger* than A. With large A in Eq. 5-76, the burning time will be small. Once turbulence enters into our problem, all conjectures become qualitative; quantitative treatment of turbulent motion in an engine cylinder is presently beyond our capabilities.

We have to conclude, then, that any simple calculations that aim at securing a value for the time of combustion can only be verified by experiment. It is tempting to argue, however, that the simple approach we have taken may have *relative* merits. For example, in the case of the dome-head combustion chamber (Fig. 5.32), since turbulence may be assumed to be identical for both spark plug positions, and since the burning velocity variations should be closely similar, the *ratio* of calculated burning times could be meaningful and relevant. At present, that speculation could only be checked by experiment.

In this section we have attempted to formulate a simple strategy for estimating the time required for combustion. The problem is an important one, for we have seen that power output is sensitive to combustion time, $\Delta\theta_c$. When we examine the burning time calculation, we find that we cannot go very far without introducing assumptions that move the mathematical model far away from reality. And there we leave the matter.

5.11 THE EXHAUST AND INTAKE PROCESSES

The analyses of the exhaust and intake processes for full throttle and part throttle operation (Sections 5.5 and 5.6) were based on the assumption of a piston moving very slowly. Pistons and rotors move at finite speeds, which can become large; when large, their motions exert a controlling influence on engine performance. In this section the exhaust and intake strokes are reexamined, and new schemes for following the progress of events are developed. These schemes require computer programs that are somewhat involved in comparison to other processes in the engine cycle. However, the effort of following through the analysis and translating the results into programs, is well rewarded, for the calculated performance curves begin to duplicate real engine operation.

Figure 5.33 illustrates the exhaust and intake processes, and a control volume analysis fits the problem easily. Consider the exhaust stroke first in Fig. 5.33a, which is drawn for a reciprocating piston engine. The results obtained for that engine apply without change to a rotary engine, although a few minor points in the analysis for the

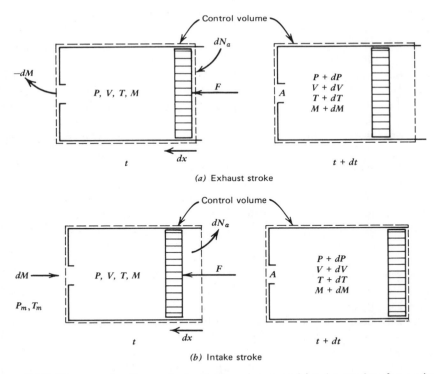

Figure 5.33 Diagrams for the analysis of the exhaust and intake strokes for a piston moving at finite velocity.

rotary engine might appear in different form. The piston moves from right to left. At some engine volume V_x, the pressure and temperature, P_x and T_x, are known, and the mass of exhaust gas M_x is known also. At some time t, the state of affairs in the engine is characterized by P, V, T, and M, while at later time $t + dt$, these have changed to $P + dP$, $V + dV$, $T + dT$, and $M + dM$. Conservation of mass requires that $-dM$ leave the cylinder in the interval dt. (Clearly, M decreases, so that dM is negative.) Let P_o denote exhaust pipe and crankcase pressure. At time t the energy in the control volume is

$$Mc_v T \tag{a}$$

During time interval dt, energy

$$- dMc_p T \tag{b}$$

leaves the control volume, and work done by the force F enters the control volume,

$$F \, dx = (P - P_o) A_p \, dx = - (P - P_o) \, dV \tag{c}$$

where A_p is the piston area and dx is positive as drawn in the figure. During dt, some air, dN_a, will enter the control volume, bringing in energy

$$h_a \, dN_a = (u_a + P_o v_a) \, dN_a = u_a \, dN_a - P_o \, dV \tag{d}$$

We assume no heat transfer across the control volume boundary. The energy in the control volume at time $t + dt$ is

$$(M + dM) c_v (T + dT) + u_a \, dN_a \tag{e}$$

In terms of these quantities, a statement of energy conservation is formed from

$$(a) - (b) + (c) + (d) = (e)$$

Inserting each term, cancelling, and dropping the second-order terms that occur in expression (e), the equation reduces to

$$-P \, dV + (c_p - c_v) T \, dM - Mc_v \, dT = 0$$

The exhaust gas is considered to be an ideal gas, so

$$\frac{dP}{P} + \frac{dV}{V} = \frac{dM}{M} + \frac{dT}{T}$$

Eliminating dT between these last two equations,

$$\frac{1}{k_p}\frac{dP}{P} + \frac{dV}{V} - \frac{dM}{M} = 0, \qquad k_p = \frac{C_p}{C_v} \qquad\qquad 5\text{-}82$$

Note that Eq. 5-82 may be integrated at once to give

$$P\left(\frac{V}{M}\right)^{k_p} = Pv^{k_p} = \text{constant}$$

which could have been written down at once, without going through this analysis, merely by noting that the gas that remains inside the engine undergoes an adiabatic, reversible change of state. Whether the change is caused by piston motion or by gas flow or by some combination of the two is immaterial. Since we wish to follow events as the piston moves, we write Eq. 5-82 in the working form

$$\frac{dP}{dt} = k_pP\left[\frac{1}{M}\frac{dM}{dt} - \frac{1}{V}\frac{dV}{dt}\right] \qquad \text{(exhaust)} \qquad 5\text{-}83$$

Turning to the intake stroke and proceeding in the same fashion, the energy in the control volume at time t is

$$Mc_v\,T + u_a\,dN_a$$

The energy entering with mass dM (a positive quantity) is

$$c_p\,T_m\,dM$$

where T_m is the temperature of the gas in the intake pipe. The energy that enters the control volume as a result of the action of the force F is

$$-F\,dx = -(P - P_o)A_p\,dx = -(P - P_o)\,dV$$

dx being positive as drawn in Fig. 5.33b. Air is pushed out of the control volume by piston movement and carries with it energy amounting to

$$h_a dN_a = (u_a + P_o v_a)\,dN_a = u_a dN_a + P_o\,dV$$

and the energy in the control volume at time $t + dt$ is

$$(M + dM)\,c_v\,(T + dT)$$

Combining these terms into an energy conservation statement, cancelling, and dropping second-order terms, the result is

$$\frac{c_v}{c_p}\frac{dP}{P} + \frac{c_v}{c_p}\left(\frac{R}{c_v}+1\right)\frac{dV}{V} - \frac{T_m}{T}\frac{dM}{M} = 0$$

Here c_v and R refer to gas in the cylinder, while c_p refers to gas in the intake pipe. Setting

$$\frac{c_p}{c_v} = k_r$$

the working form for the intake stroke is

$$\frac{dP}{dt} = k_r P\left[\frac{1}{M}\frac{T_m}{T}\frac{dM}{dt} - \frac{1}{V}\frac{dV}{dt}\right]$$

A more convenient form is

$$\frac{dP}{dt} = k_r\left[\frac{RT_m}{V}\frac{dM}{dt} - \frac{P}{V}\frac{dV}{dt}\right] \text{(intake)} \qquad 5\text{-}84$$

Both equations, 5-83 and 5-84, require mass flow values, dM/dt, which are to be found from the results of fluid mechanics, to wit:

When a gas at P, T with heat capacity ratio k flows through an opening with effective area A (i.e., A is the product of the actual area opening multiplied by a suitable discharge coefficient) into a region at pressure P_o, the flow must be considered in two regimes, subsonic and supersonic. The critical pressure ratio that defines the two regimes is called CRIT and depends only on the k value of the gas,

$$\text{CRIT} = \left(\frac{k+1}{2}\right)^{k/(k-1)} \qquad 5\text{-}85$$

When P/P_o is less than CRIT the flow is subsonic, and

$$\frac{dM}{dt} = AP_o\sqrt{\frac{2k}{RT(k-1)}\left(\frac{P}{P_o}\right)^{(k-1)/k}\left[\left(\frac{P}{P_o}\right)^{(k-1)/k} - 1\right]} \qquad 5\text{-}86$$

while the flow is supersonic when P/P_o exceeds CRIT, and

$$\frac{dM}{dt} = AP\sqrt{\frac{k}{RT}\left(\frac{2}{k+1}\right)^{(k+1)/(k-1)}} \qquad 5\text{-}87$$

(Note that Eq. 5-87 does not contain P_o, which is characteristic of supersonic flow; what matters is where the gas is coming from, not where it is going.) On the exhaust stroke, P is the cylinder pressure, and P_o is the exhaust pipe pressure. On the intake stroke, $P = P_m$, intake pipe pressure, and P_o is the cylinder pressure.

Two comments concerning the treatment of Eqs. 5-83 and 5-84 may be helpful. The first has to do with events at the end of the exhaust stroke. With the piston at TDC, or with the rotor at minimum volume, the mass of exhaust in the working volume is M_{tdc} and represents the residual gases that will remain in the engine during the next cycle calculation. The cylinder pressure at TDC is greater than intake manifold pressure, P_m, or at the very least equal to P_m. If there is no overlapping of the valves, then the exhaust valve will have just closed at TDC, and as the piston or rotor moves, the intake valve or port will begin to open. This volume increase will cause a reduction in pressure in the working space, and at the same time, exhaust gas will flow back into the intake manifold, displacing fresh charge. This is identified in Fig. 5.34 as region A. The outward flow of exhaust will continue until P is reduced to P_m. With further piston or rotor motion, P will fall below P_m, and the flow direction will reverse. Exhaust gas which first flowed into the intake manifold will return to the working volume, as in region B.

Fresh charge will not begin to enter the cylinder until all the exhaust has been returned to the working space. In the correct treatment of the intake stroke, it is necessary to identify the instant when $M = M_{tdc}$. That condition marks the start of fresh charge flow into the cylinder. The purpose of a detailed treatment of the intake stroke is an accurate computation of the mass of fresh charge in the engine at the start of the compression process, and this is achieved by summing the small amounts of

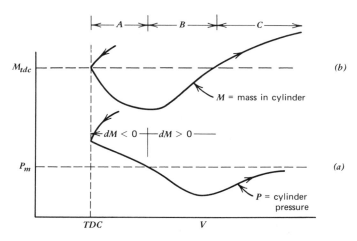

Figure 5.34 Details of events early in the intake stroke.

fresh charge, dM, that flow in during each step in the numerical integration. If the summation is to be accurate, we must know when to commence the summing process.

The second comment has to do with a suggested technique for integrating Eqs. 5-83 and 5-84. When we let

$$Q = k_p P \left[\frac{1}{M} \frac{dM}{dt} - \frac{1}{V} \frac{dV}{dt} \right] \quad \text{(exhaust stroke)}$$

$$Q = k_r \left[\frac{RT_m}{V} \frac{dM}{dt} - \frac{P}{V} \frac{dV}{dt} \right] \quad \text{(intake stroke)} \qquad 5\text{-}88$$

Eqs. 5-83 and 5-84 take the form

$$\Delta P = Q \, \Delta t \qquad 5\text{-}89$$

Now define for the exhaust stroke and for the intake stroke a maximum allowable pressure change for every step in the integration; that is,

$$\Delta P_{max} = \frac{P - P_o}{m} \quad \text{(exhaust stroke),} \qquad \Delta P_{max} = \frac{P_m - P}{m} \quad \text{(intake stroke)}$$

where m is a number between, say, 3 and 10. Also, define a maximum allowable time interval for each step in the integration

$$\Delta t_{max} = \frac{n}{360} \frac{60}{\text{rpm}} \text{(sec)}$$

that is, Δt_{max} is the time required for n degrees of crankshaft or rotor rotation. We could choose n between, again, 3 and 10.

In order to make the next step in either the exhaust or the intake strokes, first calculate ΔP_{max}, for the value of P (cylinder pressure) that prevails. With this ΔP_{max} we can calculate, from Eqs. 5-86 and 5-87, the maximum value for dM/dt, and since engine or rotor rotation rate fixes dV/dt, we have a value for Q from Eqs. 5-88 and 5-89, whichever applies. The effects of gas flow and piston or rotor movement oppose each other during both strokes of the scavenging process, so for ΔP_{max} and the operating speed which is given, Q may be positive, negative, or zero. Since increments in time must be positive, ΔP from Eq. 5-89 has the sign of Q. Then, with a value for Q, we can compare ΔP_{max} and Δt_{max}, through Eq. 5-89, and use the smaller of the two as the determinant for that step in the integration process. Figure 5.35, a flow chart of the logic that controls each step in the integrations, may be more informative than a verbal description.

Some form of control must oversee the integration, for if a uniform Δt is used at

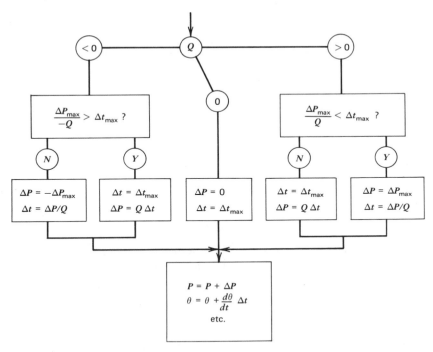

Figure 5.35 Flow chart for routing traffic during the integration of Eqs. 5-83 and 5-84. See Eq. 5-89.

each step, it may lead to ΔP values that reduce cylinder pressure below P_o during the exhaust process or above P_m on the intake stroke. On the other hand, with a fixed ΔP value, excessively large Δt values may result, overshooting the limits of piston or rotor travel.

The scheme described above will work satisfactorily for the entire exhaust stroke and for that portion of the intake stroke identified as region C in Fig. 5.34. But it will not work during the early period of the intake stroke, because if we limit pressure changes to

$$\Delta P_{\max} = \frac{P - P_o}{m}$$

P cannot drop below P_o, or, to put the matter another way, we cannot use the relation

$$\Delta P_{\max} = \frac{P_m - P}{m}$$

for the intake stroke until P becomes less than P_m. Once TDC is reached on the exhaust stroke, the procedure is to decrease P repeatedly by small prescribed amounts, using Eq. 5-89 to compute the corresponding Δt, until P becomes less than P_m (region A in Fig. 5.34), after which intake of gas from the intake pipe will begin, and Fig. 5.34 can be applied. Note, of course, that the first pieces of gas that flow in (region B in Fig. 5.34) will be returning exhaust that flowed out of the cylinder during A.

It is common, if not universal, practice to overlap closing of the exhaust valve and opening of the intake valve at TDC. This results, for the most part, in improved scavenging and increased power. The elementary theory discussed here can be applied to valve overlap, although the calculations in region A, Fig. 5.34, become more involved. With cylinder pressure P greater than exhaust pipe pressure P_o and greater than intake pipe pressure P_m, when both valves are open, exhaust gas will flow out of both valves. It is necessary to keep track of that portion of exhaust which flows into the intake pipe, since it will return before fresh charge can enter. If the exhaust valve remains open for a long period, exhaust may flow from the exhaust pipe, through the cylinder, and into the intake pipe. When this occurs, engine performance will deteriorate.

A complete treatment of the scavenging process would take into account the velocity distribution of gas in the cylinder, since its momentum affects the flow of gas out of the cylinder. Likewise, once flow has begun to carry fresh charge into the cylinder, its velocity along the intake pipe contributes to the flow rate. An analysis that includes these momentum terms is considerably more complicated and will not be examined beyond these qualitative remarks.

A meaningful definition and formulation for the pumping efficiency or *volumetric efficiency* can now be stated as the ratio of the volume occupied by the fresh charge which is drawn into the engine on the intake stroke, measured at intake pipe conditions, and divided by the displacement volume. The mass of gas in the cylinder at TDC, M_{tdc}, is known. By summing the dM amounts of fresh charge that enter during region C, Fig. 5.34, the mass in the cylinder at BDC, M_{bdc}, is known,

$$M_{bdc} = M_{tdc} + \Sigma \, dM$$

Then

$$\eta_v = \text{volumetric efficiency} = \frac{(M_{bdc} - M_{tdc})RT_m}{P_m \text{VDISP}} \qquad \text{5-90}$$

In order to carry out cycle calculations that incorporate the detailed treatment of the exhaust and intake processes, we need to specify the variation of A, the effective valve area that appears in Eqs. 5-86 and 5-87, with crank angle or with rotor position. A simple scheme, which will be used here, is

Exhaust valve: $A = AO(\sin \theta)^{1/3}$;   Opens at BDC, $\theta = 0$
                                                    Closes at TDC, $\theta = 180°$

Intake valve: $A = AO(|\sin \theta|)^{1/3}$;   Opens at TDC, $\theta = 180°$ 5-91
                                                    Closes at BDC, $\theta = 360°$

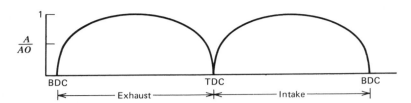

Figure 5.36 Variation of valve area A with crank angle, Eqs. 5-91.

where AO denotes the wide-open value for A and will be assumed to be identical for both valves. Generally, the exhaust valve area is larger than intake valve area. That feature can be made part of the computer program, of course. For the expressions shown in Eqs. 5-91, Fig. 5.36 shows A/AO against crank angle.

The final item required to complete a cycle calculation is WLOOP, the work associated with the exhaust and intake strokes.

WLOOP = work done on the gases during the exhaust and intake strokes

$$= - \Sigma \left(P + \frac{\Delta P}{2} \right) \Delta V$$

where ΔP denotes the change in pressure that accompanies the volume change ΔV. The summation extends from the point in the cycle at which the exhaust valve opens to the point at which the intake valve closes.

In the balance of this section, Tables 5.8 to 5.14 have been included to demonstrate how power output and other engine parameters are affected by progressive burning and by the exhaust and intake strokes and to provide the reader with sets of numbers against which program results may be checked.

Unless specified otherwise, the tables and figures apply to the following engine:

Bore = 10.16 cm   (VDISP = 0.823 liter)
Stroke = bore
Connecting rod length = 2 x bore
Compression ratio = 9:1
Full throttle (intake manifold pressure = 1 atm)
Intake manifold temperature = 300 K   5-92
Chemically correct $C_8 H_{18}$/air mixture
Valve timing per Eq. 5-91
Wide-open valve areas = AO = 3.61 cm^2
Duration of combustion = $\Delta\theta_c$, Eq. 5-75
Rate of combustion, Eq. 5-72
Spark advance = $\Delta\theta_c/2$

In addition, Figs. 5.37 and 5.38 illustrate how cylinder pressure varies with volume during the exhaust and intake strokes, revealing that the work associated with the scavenging process may not be insignificant and revealing also the profound effect the scavenging process exerts in the way of limiting power output.

In Table 5.8, as expected, as AO increases, so does power, owing to the improvement in volumetric efficiency. Surprisingly, with low volumetric efficiency of 0.30, the fresh charge accounts for 87% of the gases in the cylinder at the start of the compression stroke. (The 87% is on a mole basis).

Figure 5.37 illustrates pressure against volume during the exhaust and intake strokes for two AO-values. At 3000 rpm, with $AO = 4.51$ cm^2, pressure falls steadily after the exhaust valve opens at BDC, whereas with the small AO, pressure rises slightly toward the end of the exhaust stroke as the result of the competing effects of

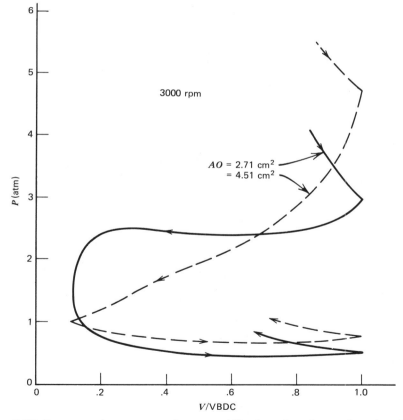

Figure 5.37 Pressure-volume traces for two **AO**-values for the engine described in Eq. 5-92.

TABLE 5.8
INFLUENCE OF AO, WIDE-OPEN VALVE AREA, EQS. 5–91, ON POWER.
(ENGINE DESCRIBED IN EQS. 5–92 RUNNING AT 3000 rpm)

AO (cm^2)	Power		η_v	η_{th}	$\dfrac{NM}{NM + NX}$	$T1$ (K)
	hp	kW				
1.81	7.2	5.4	0.30	27.7	0.87	394
2.71	14.4	10.7	0.50	30.9	0.94	345
3.61	21.9	16.3	0.68	34.2	0.96	335
4.51	27.3	20.4	0.79	36.5	0.97	335
5.41	31.2	23.3	0.87	37.9	0.97	334

piston movement and flow out of the cylinder. The other feature displayed in Fig. 5.37 is the difference in throttling effects of these two AO values on the intake stroke. With the smaller valve area, pressure at the end of the intake stroke reaches just slightly over 0.5 atm, as compared to just under 0.8 atm with the large valve area. The pressure (and the temperature) at the start of the compression stroke determine, in part, the quantity of combustible mixture in the cylinder, which in turn controls the power developed.

In addition to changing AO in Eqs. 5-91, the expression itself may be altered. This amounts to changing the profile of the valve cam in a piston engine or changing the shape and position of the ports in the bore of a rotary engine. In piston engines the most desirable combination is quick-opening, high-lift valves. That combination will ensure best scavenging of exhaust and largest intake of fresh charge. As for the thermodynamic analysis, any imagined valve arrangement is permitted and may be examined for its effect on power. In practice, this is not at all the case. There are practical limits to the rapidity with which a valve may be opened or closed. If either is excessive, the cam follower will lose contact with the cam, resulting in bouncing and, very probably, failure of some piece in the valve mechanism. By contrast, there are no moving parts, except the rotor itself, in a rotary engine, so the mechanics of port opening and closing is quite simple. Furthermore, the ports can be shaped in many ways, limited only by consideration of wear on the seals located at each apex of the rotor. It is a fact that simplicity of porting in the rotary engine, as well as the power limitations that valving exerts in piston engines, has been a key factor and strong encouragement toward development of the rotary engine. And there is evidence to support the superiority of the rotary engine; it does not display the falling off of power at high speed that characterizes all piston engines (see Fig. 5.10).

Note, in Table 5.9, how $T1$, temperture at the start of the compression stroke, increases as compression ratio decreases. This high temperature reduces the amount of fresh charge in the cylinder, accounting for a part of the power decrease at low compression ratio. The surprise in Table 5.9 is the virtually constant volumetric

TABLE 5.9
INFLUENCE OF COMPRESSION RATIO, *CR*, ON POWER
(ENGINE DESCRIBED IN EQS. 5–92 RUNNING AT 3000 rpm)

CR	Power		η_v	η_{th}	$\dfrac{NM}{NM + NX}$	T1 (K)
	hp	kW				
4	13.7	10.2	0.64	23.3	0.91	399
5	16.2	12.1	0.65	26.5	0.93	373
6	17.9	13.4	0.65	29.1	0.94	360
7	19.6	14.6	0.66	31.1	0.95	350
8	21.0	15.7	0.67	32.9	0.96	340
9	21.9	16.3	0.68	34.2	0.96	335

efficiency. The thermal efficiencies in the table can be compared with the expectations that follow from Eq. 5-8.

In Table 5.10, the engine reaches peak power at 2500 rpm. This speed would be increased if, for example, the wide-open valve area were to be increased.

Figure 5.38 illustrates pressure-volume traces for two speeds for the engine examined in Table 5.10. At 1000 rpm, pressure drops rapidly after the exhaust valve opens at BDC and remains close to 1 atm during the remainder of the exhaust stroke and the subsequent intake stroke. At a very high speed, 5000 rpm, pressure rises on approach to TDC, revealing how flow and piston movement compete. On the intake stroke, pressure drops as low as 0.4 atm, reaching 0.45 atm at the start of compression. The small arrow (↑) in the figure indicates the point on the intake stroke at 5000 rpm when fresh charge begins to flow into the cylinder from the intake manifold. Between TDC and the arrow point, exhaust flows into the intake manifold and then returns to the cylinder. The arrow position is about 40° crankshaft rotation after TDC.

TABLE 5.10
VARIATION OF POWER WITH SPEED (ENGINE DESCRIBED IN EQS. 5–92)

rpm	Power		η_v	η_{th}	$\dfrac{NM}{NM + NX}$	T1 (K)
	hp	kW				
1000	13.2	9.85	0.99	42.2	0.97	330
1500	18.3	13.7	0.95	40.7	0.97	334
2000	21.3	15.9	0.87	38.8	0.97	334
2500	22.1	16.5	0.77	36.6	0.96	335
3000	21.9	16.3	0.68	34.2	0.96	335
3500	20.4	15.2	0.58	31.9	0.95	337
4000	18.5	13.8	0.50	29.9	0.94	346
4500	16.6	12.4	0.43	28.4	0.93	361
5000	15.2	11.3	0.38	26.7	0.91	365

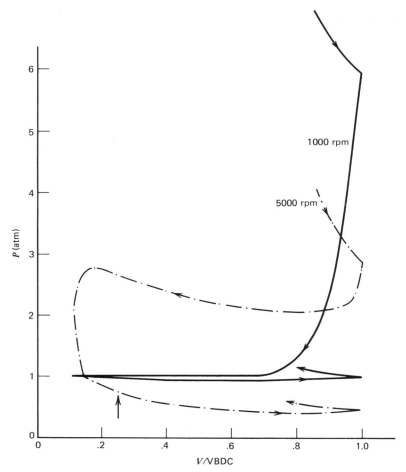

Figure 5.38 Pressure-volume traces for the engine in Eq. 5-92, operating at two speeds.

As mentioned above, and as illustrated in Table 5.8, the wide-open valve area may be varied, and the shape of the valve area curve against crankshaft angle may be altered, disregarding all the complexities of mechanical motion of the valve train mechanism. In addition, the points in the cycle when valves open and close may be altered. The effects of changes in valve timing can be examined and is shown in the next three tables.

Let θ_{ex} denote crankshaft angle when the exhaust valve begins to open, and let θ_{in} denote crankshaft angle when the intake valve has closed. Then the effective areas, A_{ex} and A_{in} presented by the valves, in terms of crankshaft angle θ (with $\theta = 0$ at

BDC), can be expressed as

$$A_{ex} = AO \sin \left\{ \left| 180 \frac{\theta + \theta_{ex}}{180 + \theta_{ex}} \right| \right\}^{1/3}$$

$$A_{in} = AO \sin \left\{ \left| 180 \frac{\theta - \theta_{in}}{180 - \theta_{in}} \right| \right\}^{1/3}$$

5-93

These expressions are chosen for simplicity, since they reduce to Eqs. 5-91 for $\theta_{ex} = 0$ and $\theta_{in} = 360$.

As shown in Table 5.11a, valve timing clearly has enormous effect on power. The reasons are made evident by Table 5.11b.

The numbers in parentheses in Table 5.11b refer to pressure in atmospheres. Note that the engine is running at full throttle, with 1 atm pressure in the intake manifold and that for some conditions of valve timing and speed, the pressure in the cylinder at the start of compression exceeds manifold pressure. Once BDC has been passed on the intake stroke, piston motion tends to increase cylinder pressure, opposed by flow out of the cylinder back to the intake manifold. Which of the two prevails depends on crank angle and speed, as well as on valve cam profile, Eqs. 5-93.

The analysis of valve timing effects on power and other engine parameters may be carried a step further by altering exhaust valve opening and intake valve closing independently, as shown in Table 5.12. Table 5.12 shows, among other things, that the

TABLE 5.11a
EFFECT OF VALVE TIMING ON POWER (hp)
(ENGINE DESCRIBED IN EQS. 5–92, EXCEPT FOR VALVE TIMING, WHICH IS GIVEN BY EQS. 5–93)

	Speed (rpm)				
	1000	2000	3000	4000	5000
Ex opens at BDC In closes at BDC	13.1	21.3	21.9	18.5	15.2
Ex opens 15° BBDC* In closes 15° ABDC	13.2	23.3	25.2	21.6	17.0
Ex opens 30° BBDC In closes 30° ABDC	13.1	24.1	27.3	24.8	20.1
Ex opens 45° BBDC In closes 45° ABDC	12.6	24.0	29.4	28.9	23.7
Ex opens 60° BBDC In closes 60° ABDC	11.7	22.7	29.9	29.1	25.1

*BBDC = before bottom dead center; ABDC = after bottom dead center.

TABLE 5.11b
EFFECTS OF VALVE TIMING ON VOLUMETRIC EFFICIENCY
AND ON $P1$, PRESSURE AT THE START OF COMPRESSION FOR THE ENGINE IN
TABLE 5.11a

	Speed (rpm)				
	1000	2000	3000	4000	5000
Ex opens at BDC	0.99	0.87	0.68	0.50	0.38
In closes at BDC	(1.00)	(0.90)	(0.70)	(0.54)	(0.45)
Ex opens 15° BBDC	0.98	0.92	0.73	0.55	0.40
In closes 15° ABDC	(1.01)	(0.97)	(0.73)	(0.61)	(0.51)
Ex opens 30° BBDC	0.96	0.93	0.78	0.58	0.44
In closes 30° ABDC	(1.03)	(1.04)	(0.87)	(0.71)	(0.58)
Ex opens 45° BBDC	0.92	0.90	0.81	0.63	0.48
In opens 45° ABDC	(1.06)	(1.10)	(0.99)	(0.84)	(0.69)
Ex opens 60° BBDC	0.86	0.86	0.81	0.66	0.50
In closes 60° ABDC	(1.10)	(1.16)	(1.13)	(0.94)	(0.81)

TABLE 5.12
EFFECTS OF VALVE TIMING ON POWER (hp)
(SAME ENGINE AS TABLE 5.11a)

		2000 rpm				
		Intake valve closes, degrees ABDC				
Exhaust valve opens, degrees BBDC		0	15	30	45	60
	0	21.3	22.5	22.8	22.4	21.2
	20	22.2	23.5	23.7	23.3	22.1
	40	22.7	24.1	24.3	23.9	22.6
	60	22.8	24.2	24.4	24.0	22.7
	80	22.3	23.6	23.9	23.5	22.2

		3000 rpm				
		Intake valve closes degrees ABDC				
Exhaust valve opens, degrees BBDC		0	15	30	45	60
	0	21.9	24.0	25.0	25.8	25.8
	20	23.3	25.6	26.6	27.7	27.6
	40	24.4	26.8	27.9	29.1	29.0
	60	25.0	27.5	28.7	30.0	29.9
	80	24.9	27.5	28.7	30.1	29.9

combination of valve events that produces maximum power for a given speed depends on speed.

In Tables 5.11a, 5.11b, and 5.12, intake valve opens at TDC and exhaust valve closes at TDC. In real engines, the intake valve is set to open before TDC, while the exhaust valve closes after TDC. The period when both are open is referred to as valve overlap. The overlap period may be modeled using the principles of thermodynamics and fluid mechanics described in this section. Overlap tends to require a somewhat more complicated analysis.

In a naturally aspirated engine (i.e., unsupercharged), intake manifold pressure is always less than, or in the limit equal to, exhaust manifold pressure. On approach to TDC the intake valve opens, and with the exhaust valve open, exhaust will flow from the cylinder to the intake manifold. If cylinder pressure drops below exhaust manifold pressure, exhaust will flow from exhaust manifold to cylinder as more exhaust flows from cylinder to intake manifold. The result will be a decrease in volumetric efficiency and less power than would be obtained without valve overlap. Why, then, are automotive engines manufactured with overlap? The answer is that under some conditions of operation, valve overlap does increase power, but the elementary analysis of the overlap period is too simplified to reveal that increase. For one thing, the velocity of the gases in the cylinder is ignored. For another, velocities in both manifold systems are ignored. Periodic opening and closing of valves produces intermittent motion in both manifolds, which is characterized by a certain amount of sloshing back-and-forth. By "tuning" both manifolds, which means choosing geometry carefully, volumetric efficiency can be improved with valve overlap at some speeds. (It may deteriorate at other speeds; the choice of overlap dimensions is made with considerations given to the conditions under which the motor will be operated for most of its life.)

In supercharged engines, intake manifold pressure should exceed exhaust manifold pressure. With overlapped valves, fresh charge may pass through the cylinder and out the exhaust valve and never burn. From the point of view of the owner, that fuel has been consumed and must be figured in the calculation of thermal efficiency as fuel burned.

With regard to computational schemes with arbitrary valve timing, early exhaust valve opening does not present any problem. This is not so for the intake process. Early in the intake stroke, cylinder pressure P drops below intake manifold pressure P_m and remains below P_m until BDC, or maximum volume in a rotary engine, is reached. At that point

$$P = P_m - \epsilon, \qquad \epsilon > 0$$

As the intake stroke continues, with intake valve or port open, P will increase owing to the combined effects of inflow of gas and piston or rotor movement. As time passes, ϵ will decrease. If the point of intake valve closing is delayed for a sufficiently long span of crankshaft rotation, P will eventually exceed P_m. To achieve this, however, "brute force" is required. That is, when ϵ becomes sufficiently small, say less than 1% of P_m, interrupt the calculation and set

$$P = P_m + \epsilon$$

This will reverse the direction of flow. The flow of fresh charge and residual exhaust out of the cylinder will reduce P, while piston or rotor motion will increase P, since BDC has been passed. These competing effects determine cylinder pressure at the instant the intake valve seats, or the intake port is covered.

Engine size exerts a strong influence on power, as is to be expected. In comparing engines of different size, power/unit volume of displacement is a convenient measure. Table 5.13 compares geometrically similar motors. In each case, bore and stroke are equal, and the connecting rod length is twice that of the bore. Wide-open valve area AO is proportional to the bore squared. Since combustion time $\Delta\theta_c$ can be assumed to be fixed, in part at least, by the distance the flame must travel from spark plug to the most remote point in the cylinder, $\Delta\theta_c$ has been taken proportional to the bore. Equation 5-75 is assumed to predict combustion time for a bore of 10.16 cm. Spark advance is, in all cases, set at one-half combustion time. Other features of the engine that are independent of size are as listed in Eqs. 5-92. Large engines (in a geometrically similar series) suffer poor volumetric efficiency and long duration combustion. Large engines therefore require AO values more nearly proportional to the bore cubed, rather than the bore squared, since the volume of exhaust to be scavenged is proportional to the bore cubed. In addition, dual spark plugs reduce combustion time. In theory, large engines can produce the same power per unit displacement volume as can small engines.

The final performance figures for the engine described in 5-92 is Table 5.14, showing how fuel/air ratio affects operation. Here, Y is the moles of oxygen/mole of fuel, Y_{cc} denoting chemically correct mixture. Note that Table 5.14 does not take

TABLE 5.13
VARIATION OF POWER/VDISP WITH ENGINE SIZE
(GEOMETRICALLY SIMILAR ENGINES, 3000 rpm)

Bore (cm)	Specific Power hp/in³	Specific Power kW/liter	$\Delta\theta_c$	η_v	$\dfrac{NX}{NM + NX}$	T1 (K)	AO (cm²)
2.54	0.827	37.6	15°	0.99	0.97	329	0.23
5.08	0.752	34.2	31°	0.95	0.97	334	0.90
7.62	0.601	27.3	46°	0.83	0.97	335	2.03
10.16	0.436	19.8	62°	0.68	0.96	335	3.61
12.7	0.298	13.6	77°	0.54	0.95	341	5.64
15.2	0.202	9.19	92°	0.43	0.92	358	8.12

TABLE 5.14
VARIATION OF POWER WITH FUEL/AIR RATIO
(3000 rpm)

Y/Y_{cc}	Power		η_{th}	$\Delta\theta_c$
	hp	kW		
0.68	17.4	13.0	19.1	83°
0.76	19.7	14.7	23.6	68°
0.84	20.9	15.6	27.5	61°
0.92	21.6	16.1	31.0	60°
1.00	21.9	16.3	34.2	62°
1.08	20.1	15.0	33.9	65°
1.16	18.4	13.7	33.4	69°
1.24	17.0	12.7	32.9	74°

into account the temperature drop caused by vaporization. Intake manifold temperature TM is assumed constant at 300 K, either by design or by accidental heating from, say, the exhaust manifold. Duration of combustion varies according to the empirical expression, Eq. 5-75. However, that expression is by no means a complete statement. For example, Fig. 5.39 illustrates in a qualitative fashion how combustion duration is

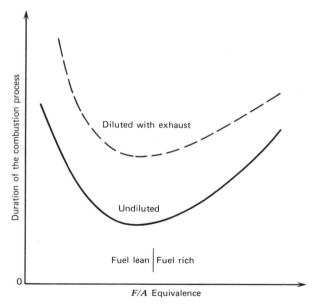

Figure 5.39 Typical variation of length of combustion process, with fuel/air mixture and with exhaust gas dilution.

affected by the amount of residual exhaust gas in the cylinder at the instant the spark is passed. When diluted with exhaust, the fuel exhibits an increase in combustion time; the inert exhaust molecules simply "get in the way" and slow down the rate of flame front advance. Quantitative data for the curves shown in Fig. 5.39 are not yet sufficient to say whether or not generally useful empirical relationships can be set down connecting $\Delta\theta_c$ and $NM/(NM + NX)$. Such relationships need to be extended well beyond the amounts of residual exhaust gas normally found in Otto cycle engines. This need arises from current experimentation with the recirculation of exhaust gas for the purpose of reducing the amount of NO_x formed by combustion. This topic is discussed briefly in the next section.

5.12 EXHAUST GAS RECIRCULATION AND STRATIFIED CHARGE

The history of the development of the reciprocating piston internal combustion engine began when the early pioneers learned how to make the thing stay together and work. Following that triumph, an enormous amount of effort has been expended to mass-produce engines with steadily increasing durability, reliability, and power. These achievements have been made possible by the developments of new materials, processes, fuels, lubricants, experimental instrumentation, and so forth.

At the present time, a considerable amount of research effort is devoted to two problems: one is the growing need to acquire control over the nature of the gases in exhaust emissions; the other is the growing need to increase thermal efficiency and reduce fuel consumption. One approach to the first problem is *exhaust gas recirculation* (EGR). The *stratified charge* (SC) engine shows excellent promise as an answer, or at least a partial answer, to the second problem. (We shall not discuss the contribution that an informed and concerned public could make through the exercise of self-restraint and forethought.) Our discussion of both problems will be qualitative; the solutions to both are still in early stages, and theories are as yet in formative stages.

A large variety of chemical species are formed during combustion of a mixture of hydrocarbon vapor, air, and residual exhaust. What species, and in what concentrations, depends on the composition, pressure, and temperature of the reactant mixture. During the expansion stroke following combustion (this discussion applies to piston and rotary engines), the concentrations change. At any time following ignition, we can calculate these concentrations, using thermodynamic methods discussed in Chapter 11. The calculated and measured concentrations may or may not agree; to apply thermodynamics, a stringent condition of equilibrium must be imposed. Recent legislation has established upper limits on the allowable concentration of certain species. One of these, NO (nitric oxide), is of particular concern, for it is a stable molecule and accumulates, since nature seems to have no use for it. Another group, hydrocarbons, HC (i.e., C_xH_y molecules), is also of concern, for their presence indicates only partial recovery of the energy in the fuel bonds and therefore a loss in efficiency and economy.

During combustion, which we have treated as a simple conversion of a reactant mixture to a product mixture, the 26 atoms in an octane molecule are not instantly set free to take part in the general chemical scramble going on. Each fuel molecule breaks up into smaller pieces, and the number of ways this can happen is, of course, extraordinarily large. Furthermore, and this seems to be the key conjecture in present theories, there appears to be widespread agreement that the flame is *quenched* as it approaches the wall of the combustion chamber, leaving a thin layer of unburned fuel. Subsequently, these unburned fuel molecules or pieces of fuel molecules may undergo further breakup, but they will not be burned, and a large fraction is swept out of the combustion chamber during the exhaust process. It would seem, therefore, that the surface area/volume ratio for the combustion chamber exerts the major control on the concentrations of HC particles. Agreement on this theory is not universal; laboratory data from which to judge have only just begun to appear. If the quenching notion is indeed correct, then it would seem to follow that chamber shape must also play a role. Since quenching is the result of heat transfer from the reactants at the wall to the wall itself, the rate of heat transfer should be larger in corners than along flat walls. The thickness of the unburned layer should therefore be larger in corners than, say, on piston faces or on rotor flanks. There is some evidence to support this conjecture; for example, HC concentration seems to be reduced by shaping the piston head corners in ways that enlarge the crevice formed by the piston, the cylinder wall, and the top piston ring.

One of the methods proposed, and which is presently undergoing extensive laboratory testing, for reducing NO concentration is exhaust gas recirculation. The procedure is to cool a portion of the exhaust and to introduce it below the carburetor, for example, thereby deliberately diluting the fresh charge. This dilution has several effects. Obviously, since the exhaust contributes no energy to the combustion process, the peak temperature reached with EGR is reduced. The formation of NO increases with increasing temperature, so that EGR works in the desired direction. The dilution reduces power, of course, and part of the reason is the longer combustion time, which is shown in Fig. 5.39. At the present stage of our discussion of internal combustion engines, the power loss associated with various percentages of EGR can be calculated readily, the calculation gaining in meaning if the ordinate in Fig. 5.39 can be given a numerical scale. The matter of calculating the benefits of EGR to NO concentration is discussed in Chapter 11. Exhaust gas is preferred as a diluent over air, owing to the fact that CO_2 and H_2O, being triatomic molecules, have higher heat capacities than O_2 and are therefore more effective in achieving lower combustion temperatures.

The stratified charge engine, or stratification, is arranged so that the fuel/air mixture in the vicinity of the spark plug is much richer than elsewhere in the combustion chamber. The combustion is "soft," since the air remote from the spark plug acts as a "cushion." With charge stratification, operation can be carried to overall fuel/air mixtures that are well below the operating limit for conventional engines, a limit set by the ignitability of the mixture — hence the gain in economy that spurs the whole development of charge stratification.

SC engines have another impressive advantage; power is controlled entirely by the rate of the fuel flow. In conventional engines with carburetors, because they operate within the ignition limits, power can only be controlled by throttling. This is a wasteful arrangement; the engine works against itself. With charge stratification the engine runs always at full throttle, so that fuel is not used wastefully. In SC engines, there must be a device limiting fuel flow, since with very rich overall mixtures, all the fuel will not burn and will appear as dense smoke in the exhaust.

The theory of stratification is exceptionally attractive; how is it achieved in practice? (Charge stratification is not a new idea, by the way. It was mentioned in the 1920s by Ricardo, one of the most prominent names in the history and development of internal combustion engines.) A popular form of engine is illustrated in Fig. 5.40. Direct cylinder head fuel injection is synchronized with piston or rotor motion. Evaporation must take place promptly, so very high pressure is imperative. Swirling air mixes with fuel carrying the mixture past the spark plug. The burning region is therefore localized and comprises a rich mixture with good ignitability in an overall lean mixture, which by itself might not burn rapidly enough to produce appreciable power. Piston heads are cupped in some engines, as shown in Fig. 5.40. The outer portion of the piston makes close approach to the cylinder head at TDC, which forces air inward during the late stages of compression. This movement is referred to as "squish." Clearly, a good deal of trial-and-error experimentation may be required to produce satisfactory stratification.

The thermodynamic analysis of stratified charge combustion must, in the idealized case, deal with what may be a large temperature difference at the conclusion of combustion. In Fig. 5.41, with P_i and T_i denoting conditions at ignition, the products will reach T_p and the cushion of air T_a with a final pressure P_f. If we ignore the differences in c_v and R values between reactants, products, and air, and if the temperature difference is dissipated at constant volume and without heat transfer,

$$M \frac{T_e}{P_e} = \frac{m_p T_p + m_a T_a}{P_f}$$

where M is the total mass

$$M = m_p + m_a = m_r + m_a$$

where subscripts a, r, and p denote air, reactants, and products. Solving for P_f

$$P_f = \frac{P_i}{T_i} \left[\frac{m_p}{M} T_p + \frac{m_a}{M} T_a \right] \, , \qquad \left(\frac{P_i}{T_i} = \frac{P_e}{T_e} \right) \qquad \text{5-94}$$

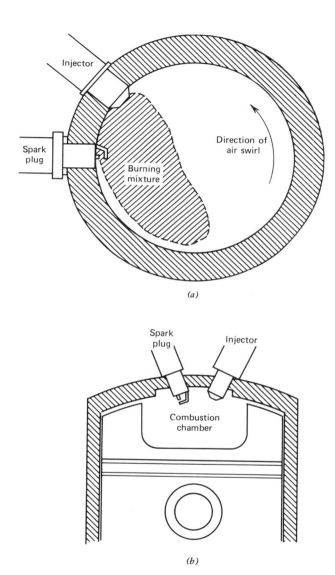

(a)

(b)

Figure 5.40 Stratified charge engine geometry. (*a*) Arrangement of injector and spark plug in a stratified charge engine. (*b*) Cupped piston in stratified charge engine, which produces "squish."

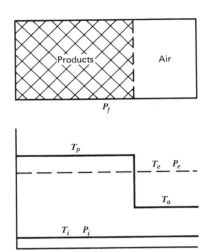

Figure 5.41 Conditions in an idealized stratified charge cylinder at the instant combustion is completed.

For an energy equation describing adiabatic combustion,

$$Mc_vT_i + m_rq = m_pc_vT_p + m_ac_vT_a$$

where q is the energy released per mole of reactant, which we can crudely approximate by

$$q = c_v(T_e - T_i)$$

Making this substitution in the energy equation and then eliminating T_p from Eq. 5-94, we arrive at a quick, but only approximate, expression for P_f,

$$P_f = P_i \left[1 + \frac{m_r}{M}\left(\frac{P_e}{P_i} - 1 \right) \right] \qquad 5\text{-}95$$

For example, suppose the overall reactants are

$$C_8H_{18} + 4.76 \times 16.5 \text{ air}$$

and the stratified charge is chemically correct

$$C_8H_{18} + 4.76 \times 12.5 \text{ air}$$

Then

$$m_r = 114 + 4.76 \times 12.5 \times 28.97 = 1837$$
$$m_a = 4.76 \times 4 \times 28.97 = 551$$
$$m \ = 551 + 1837 = 2389$$

For a 9:1 compression ratio engine, reasonable values are $P_i = 15.9$ atm, $P_e = 62.2$ atm, and $T_i = 708$ K.

Then

$$T_a = \left(\frac{P_f}{P_i}\right)^{0.285} T_i$$

since the cushion of air is isentropically compressed, and

$$P_f = P_i \left[1 + \frac{1837}{2389}\left(\frac{62.2}{15.9} - 1\right)\right] = 3.24 P_i$$

so that $T_a = 990$ K. From Eq. 5-94

$$T_p = \frac{P_f}{P_i}\frac{M}{m_p} T_i - \frac{m_a}{m_p} T_a$$

$$= 3.24 \frac{2389}{1837} 708 - \frac{551}{1837} 991 = 2685 \text{ K}$$

The temperature difference is large

$$T_p - T_a = 1695 \text{ K}$$

What is to be done with that temperature difference? Can we assume that it disappears instantly (unrealistic), allow it to remain intact during the expansion stroke (unlikely), or suggest that it dissipates according to some prescribed schedule during the compression stroke? The latter is preferred but leaves us with a schedule to be established according to whim and fancy. More experimental engine data are needed to settle that matter.

Stratification has an enormously important advantage over conventional carburetion and injection: the virtual elimination of preignition and detonation and the opportunity to burn fuel that does not require extensive refining and special additives. The topic of detonation is discussed in the next section.

5.13 DETONATION
During normal combustion, pressure rises and falls smoothly with time, as shown in Fig. 5.24. Under certain conditions, this smooth change may be interrupted by an

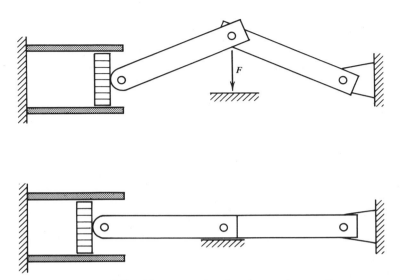

Figure 5.42 Rapid compression for detonation studies.

abrupt and large pressure change accompanied by an audible "ping" or "knock," which is familiar to motorists. This is *detonation*. In this section we examine the peak pressure that results from detonation.

Motion pictures of flame movement in combustion chambers show conclusively that detonation occurs when *all* of the remaining unburned gases ahead of the flame front ignite and burn instantaneously. The ignition of this finite mass of combustible mixture is not caused by a spark; instead, the combustion results from conditions in the unburned gas.

Figure 5.42 illustrates the type of equipment that can be used for laboratory studies of the essential features of detonation. With a linkage arrangement, a combustible mixture is rapidly compressed, and the piston remains stationary after compression is completed. The procedure consists in filling the cylinder with a combustible mixture and allowing a long period for complete mixing and uniform temperature to be reached. The mixture is then compressed rapidly, and subsequent events, if any, are observed. Under some conditions, nothing happens; under other conditions of mixture composition, initial pressure, and temperature before compression, the entire gas mixture explodes after some time interval following the end of compression. Figure 5.43 illustrates the type of data such experiments provide: a plot of temperature at the end of compression against the time interval from end of compression until detonation. Since there is no ignition source in the cylinder, detonation is self-ignition.

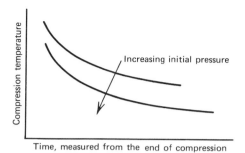

Figure 5.43 Typical temperature—time relation for detonating mixture.

When detonation occurs, the time interval may be shortened either by increasing the initial temperature before compression or by altering the compression ratio of the piston/cylinder apparatus. If the initial pressure before compression is increased, the time interval before detonation is decreased. Two important generalizations emerge from these experiments and appear to have broad applicability:

1. Below a characteristic threshold temperature, detonation will not occur.
2. When brought to a temperature higher than the threshold, the mixture will detonate if left undisturbed for a sufficient length of time.

During normal, progressive combustion the temperature of the unburned gas rises, and detonation will occur when the residence time above threshold temperature is sufficiently long.

The conditions that generally seem to favor the appearance of detonation are wide-open throttle and low engine speed. Intake manifold pressure is high at wide-open throttle, and turbulence level is low at low speed.

The results of detonation can be obtained by a simple model. Progressive combustion is assumed, for which some rate of combustion expression (Eqs. 5-70, 5-71, 5-72, etc.) is selected as representing events in the cylinder. At every step in the combustion period calculations, the pressure, P, temperature, TR, mass, and volume of the reactants are known. With this information, the temperature TP of the products of combustion (assumed to be uniform) can be computed.

In order to simulate detonation, progressive combustion may be halted at any time we choose, and then the remaining unburned reactants are assumed to burn adiabatically and at *constant volume*. The products that detonate reach pressure PD and temperature TD. As Table 5.15 shows, PD reaches horrendous values; the smaller the fraction that detonates, the larger is the PD.

At the instant detonation is completed, the combustion space contains two

TABLE 5.15
FULL THROTTLE; CHEMICALLY CORRECT C_8H_{18}/AIR; $CR = 9$;
$TM = 300$ K; VDISP = 0.823 liter; VALVE TIMING PER EQS. 5–91; $AO = 3.61$ cm^2;
RATE OF COMBUSTION, EQ. 5-70 (linear)

Fraction Detonating	P (atm)	TR (K)	TP (K)	PD (atm)	TD (K)	PE (atm)	TE (K)	Power hp	Power kW

Combustion time $\Delta\theta_c = 20°$, Spark advance = $10°$
1000 rpm

Fraction Detonating	P (atm)	TR (K)	TP (K)	PD (atm)	TD (K)	PE (atm)	TE (K)	hp	kW
20%	74	1008	3049	262	3372	88	3114	14.0	10.4
15%	77	1016	3062	270	3380	87	3110	14.0	10.4
10%	79	1025	3073	276	3386	86	3105	14.0	10.4
5%	81	1032	3082	282	3392	85	3098	14.0	10.4
None	—	—	—	—	—	83	3091	14.0	10.4

Combustion time $\Delta\theta_c = 30°$, Spark advance = $15°$
1500 rpm

20%	71	997	3030	254	3364	85	3097	20.7	15.4
15%	73	1003	3035	258	3368	83	3085	20.7	15 4
10%	74	1007	3039	262	3372	80	3072	20.7	15.4
5%	75	1010	3039	264	3374	78	3056	20.7	15.4
None	—	—	—	—	—	75	3038	20.7	15.4

parcels of product gases at different pressure and different temperature. In Table 5.15, power calculations were carried out by assuming that the pressure difference relaxed instantly to PE and that the temperature difference relaxed instantly to TE, which are calculated from conservation of energy and constant volume during the mixing process. It is quite likely that the pressure variation smoothes out rapidly in real engines, whereas the temperature variation may persist for some time and dissipate slowly. The table shows detonation to have no effect on power, which agrees with experience. This is not to say that detonation does not cause damage to piston heads and cylinder heads, for which reasons it is to be avoided.

Preignition, as the term implies, describes ignition that occurs prior to passage of the spark at the plug and is invariably caused by a local hot spot. Preignition is very different from detonation, although there is frequently a connection between the two. Preignition can promote detonation by increasing pressure and temperature throughout the unburned gas, while detonation can lead to formation of a localized hot spot.

5.14 AIR STANDARD DIESEL CYCLE

The air standard Diesel cycle is drawn in Fig. 5.44, approximately to scale, for a compression ratio $V2/V1 = 14$, and a volume ratio $V3/V2 = 3.9$. The four parts of the

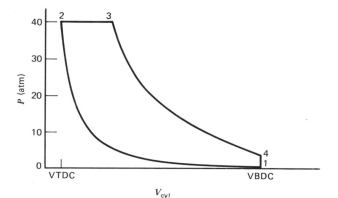

Figure 5.44 Air standard Diesel cycle. CR = 14, V3/V2 = 3.9.

cycle are

1–2	adiabatic compression
2–3	constant pressure heating
3–4	adiabatic expansion
4–1	constant volume cooling

The working fluid, an ideal gas, remains in the engine and performs a cycle.

The thermal efficiency can be formulated as

$$\eta_{th} = 1 - \frac{\text{heat rejected}}{\text{heat added}}$$

$$= 1 - \frac{CR(T4/T1 - 1)}{k(T3/T2 - 1)} \qquad \text{5-96}$$

$$= 1 - \frac{VR^k - 1}{kCR^{k-1}(VR - 1)}$$

where $CR = V2/V1$, $VR = V3/V2$, and $k = c_p/c_v$.

Whereas Otto cycle efficiency depends only on the compression ratio (Eq. 5-8), Diesel cycle efficiency requires specifying the volume ratio associated with the heating process as well as the compression ratio.

5.15 COMBUSTION IN DIESEL ENGINES

In real Diesel engines, the heating process between points 2 and 3 of Fig. 5.44 is replaced by combustion. High compression ratios bring the air to a high temperature at the end of the compression stroke. Fuel injection begins at once, and the quantity of fuel injected is synchronized with piston movement to keep cylinder pressure constant. Combustion begins as soon as injected fuel makes contact with hot air and continues in that manner, without need for a spark. Diesel engines are sometimes referred to as compression ignition engines (CI).

The object now is to calculate the net work done in a single cycle. We shall do this for a Diesel engine operating at full throttle. The work associated with the compression stroke is

$$\text{WCOMP} = (NA + NX)CVR \ CVR \ (T2 - T1)$$

where NA and NX denote the quantities of air and residual exhaust, CVR is the constant volume heat capacity of that mixture, while $T1$ and $T2$ are the temperatures at the start and the end of the compression stroke. Since NX and $T1$ are influenced by the previous cycle, we have to assume values for both in order to get the calculations underway. The assumed values are not critical; the cycle will converge even for ridiculous NX and $T1$ values.

For constant pressure adiabatic combustion, the energy equation is

$$H_r(T2) + H_f \ (T_f) = H_p(T3)$$

where subscripts r, f, and p refer, respectively, to the reactants $NA + NX$, the fuel, and the products of combustion. That equation may be rewritten

$$Q_p = H_p(T3) - H_p(T2) \qquad \qquad 5\text{-}97$$

where Q_p is defined in Eq. 2-14. Equation 5-97 provides the value for $T3$, temperature at the conclusion of combustion, with which we can now find $V3$,

$$V3 = \frac{NP \ R \ T3}{P2} \qquad \qquad 5\text{-}98a$$

and then $T4$ the temperature at the end of adiabatic expansion

$$T4 = T3 \left(\frac{V3}{\text{VBDC}}\right)^{KP \ - \ 1} \qquad \qquad 5\text{-}98b$$

The work accomplished on the expansion stroke is

$$\text{WEXP} = P2(V3 - V2) + U_p(T3) - U_p(T4) \qquad \qquad 5\text{-}99$$

The exhaust/intake strokes can now be passed through, assuming very slow piston motion, as in Section 5.5. This leads, with Eq. 5-29, to the $T1$ value for the ensuing cycle. Calculations continue until successive cycles exhibit identical numbers.

Table 5.16 illustrates the type of results that will be obtained using a 14:1 compression ratio, with $C_{10}H_{22}$ chosen to represent a typical Diesel fuel. Cycle convergence is obtained on the third cycle. The residual exhaust amounts to only 2% on a mole basis.

TABLE 5.16
IDEAL DIESEL CYCLE; FULL THROTTLE; CHEMICALLY CORRECT $C_{10}H_{22}$/AIR; CR = 14; VDISP = 0.823 liter; TM = 300 K; 3000 rpm

Cycle	Power		$P2$	$T1$	$T2$	$T3$	$V3/V2$	$\dfrac{NA}{NA + NX}$	η_{th}
	hp	kW	(atm)	(K)	(K)	(K)			
1	42.6	31.8	40.0	300	853	2832	3.54	1.00	41.6
2	40.2	30.0	39.9	318	903	2845	3.36	0.98	42.3
3	40.2	30.0	39.9	318	903	2845	3.36	0.98	42.3

We want to examine Diesel engine combustion in greater detail, for reasons that will become evident shortly. In order to achieve, or to approximate, constant pressure during the combustion process, fuel must be injected and burned over a finite span of crankshaft rotation. Diesel combustion is progressive, and the results obtained in Section 5.9 may be applied here. If mass fraction, Δn, of fuel burns while the piston moves through a volume change amounting to ΔV, the resulting change in pressure, ΔP, will be given by Eq. 5-67, repeated here in slightly different terms, namely,

$$\Delta P = - PK \frac{\Delta V}{V} + (P3' - P2) \frac{\text{VTDC}}{V} \Delta n \qquad \text{5-100}$$

where $K = c_p/c_v$,
   $P2$ = pressure at TDC, at end of the compression stroke,
   $P3'$ = the pressure that would be reached if *all* the fuel burned instantaneously at TDC.

When we set $\Delta P = 0$ in Eq. 5-100, along with $P = P2$, and assume that K remains constant during combustion, Eq. 5-100 then integrates to

$$n = \frac{V - \text{VTDC}}{V3 - \text{VTDC}} \qquad \text{5-101}$$

where $V3$ is the volume at the instant combustion is completed and can be found from

$$V3 = \text{VTDC} \left[1 + \frac{P3' - P2}{K\, P2} \right] \qquad\qquad 5\text{-}102$$

Equation 5-101 describes the manner in which fuel must be injected and then burned if constant pressure is to be maintained. Thermodynamics makes no distinction between injection and burning, whereas in real engines, there could be a time lag between injection and burning.

Now examine the time rate of burning, represented by the ratio dn/dt. Writing the quotient in the following form:

$$\frac{dn}{dt} = \frac{dn}{dV}\frac{dV}{d\theta}\frac{d\theta}{dt}$$

$$= \frac{1}{V3 - \text{VTDC}} \cdot V'(\theta)\frac{d\theta}{dt} \qquad\qquad 5\text{-}103$$

Here, θ denotes engine crank angle, while $V'(\theta)$ represents the functional relation between engine volume and θ. The first term on the right in Eq. 5-103 comes from Eq. 5-101; that is, we assume we are dealing with an engine designed for constant pressure combustion, starting at TDC. The total quantity of fuel to be burned fixes $V3$, so the first term in Eq. 5-103 may be treated as a constant. According to Eq. 5-103, the time rate of burning required to maintain constant pressure is proportional to engine speed and increases with θ as burning proceeds, since $V'(\theta)$ is a positive function. Figure 5.45 illustrates the meaning of Eq. 5-103.

In the figure, we have included what seems to be an altogether reasonable expectation, namely, that there is an upper limit to the allowable and achievable burning

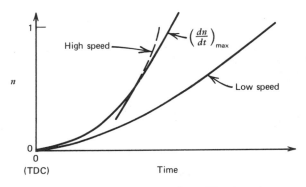

Figure 5.45 Burning curves for a Diesel engine.

rate, in time, which cannot be exceeded. As indicated in Fig. 5.45, it is entirely reasonable to suppose that engine speed could be raised to a value such that the maximum allowable burning rate is reached *before* all the fuel has been injected and burned. If this condition is found to exist, then we assume that all subsequent burning would proceed at the maximum burning rate, not at the burning rate required for constant pressure combustion, as in Eq. 5-103. We can expect, therefore, that if the maximum burning rate is reached, the balance of the combustion process will not maintain a constant pressure; instead, pressure will fall off and with it a decrease in work output for the cycle.

The detailed treatment of Diesel combustion can now be outlined. We begin with known values for $P2$ and $T2$, conditions at the end of the compression stroke. The fuel/air ratio is known. We now compute $P3'$, the pressure following *constant volume* combustion at top dead center, and then $V3$, from Eq. 5-102, the volume at the end of *constant pressure* combustion.

Suppose the maximum burning rate is known or some guess value is agreed upon. Find $\theta 3$, the crank angle corresponding to $V3$, and for the given speed, compute

$$\left(\frac{dn}{dt}\right)_{\theta=\theta 3}$$

from Eq. 5-103. As indicated in Fig. 5.45, the maximum required burning rate occurs at the end of combustion. We should compare it with the maximum allowable burning rate. If the maximum allowable burning rate is not exceeded, then we know the entire combustion process will proceed at constant pressure. We have $T3$ from Eq. 5-98a and the work done on the expansion stroke from Eq. 5-99.

On the other hand, if the burning rate calculation indicates that the maximum will be exceeded, then the combustion process must be followed in stepwise fashion. Suppose we choose to carry of combustion in N steps, so that

$$\Delta\theta = \frac{\theta 3 - 180}{N}$$

$$\Delta V = V'(\theta)\,\Delta\theta$$

and as long as the maximum burning rate is not exceeded,

$$\Delta n = \frac{\Delta V}{V3 - VTDC} \qquad\qquad 5\text{-}104$$

Once the burning rate reaches the maximum allowable, in place of Eq. 5-104

$$\Delta n = \left(\frac{dn}{dt}\right)_{max}\Delta t \qquad\qquad 5\text{-}105$$

for all subsequent increments in time. The integration is completed (i.e., combustion is completed) when

$$\Sigma \, \Delta n = 1 \qquad\qquad 5\text{-}106$$

and the work of expansion is calculated from

$$\text{WEXP} = \Sigma \left(P + \frac{\Delta P}{2} \right) \Delta V + U_p(T3) - U_p(T4) \qquad\qquad 5\text{-}107$$

where the summation applies for the entire combustion process, and $T3$ is the temperature at the end of combustion.

Figure 5.46 illustrates how the pressure-volume trace of the Diesel cycle is altered when the maximum allowable burning rate is reached. When the maximum rate is large, say 900 1/s, it is reached, in this case, just before the end of combustion, whereas a rate of 500 1/s is reached early in the combustion process, followed by a sharp decrease in pressure and in power output.

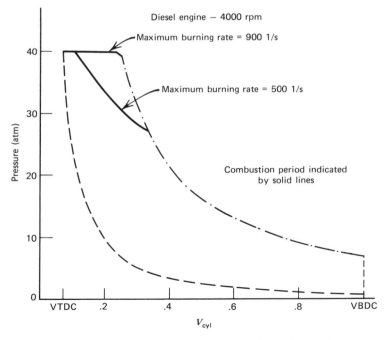

Figure 5.46 Pressure volume traces for two assumed maximum burning rates. For description of engine, see Table 5.16. Fuel injection rate per Eq. 5-101.

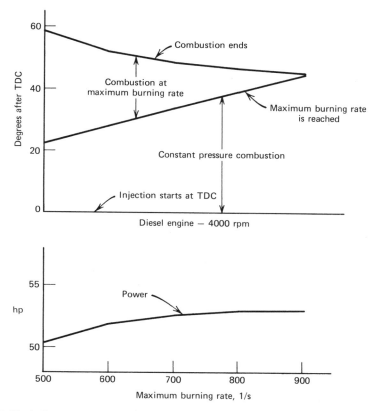

Figure 5.47 Influence of maximum burning rate. For description of engine, see Table 5.16. Fuel injection rate per Eq. 5-101.

Figure 5.47 is another way of looking at the same calculations. Injection and combustion begin at TDC. As the maximum allowable burning rate increases, it is reached later and later in the combustion process, and the total crankshaft span of rotation for combustion decreases.

Injection may be initiated at any crank angle of our choosing. If θ_i denotes crank angle at the start of injection, then in place of Eq. 5-103,

$$\Delta n = \frac{V'(\theta - \theta_i)}{V3 - \text{VTDC}} \Delta\theta$$

5-108

Figure 5.48 illustrates just how drastically the pressure-volume trace of the cycle can change as the start of injection is changed. With early injection, $10°$ before TDC,

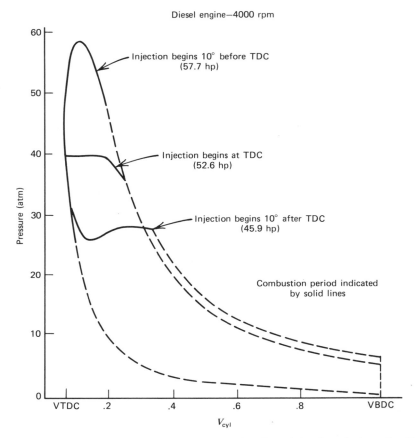

Figure 5.48 Diesel engine. Pressure-volume traces for three injection timings. See Table 5.16 for description of engine. Fuel injection rate per Eq. 5-101.

there is a substantial gain in power but at the expense of an equally substantial increase in peak pressure. On the other hand, by delaying injection until $10°$ after TDC, peak pressure is reduced considerably, but so also is power. The sharp changes in the combustion curves with $10°$ before TDC, and TDC injection, indicate that the maximum allowable burning rate has been reached.

Unfortunately, thermodynamics does not provide us with the slightest shred of information regarding probable values for the maximum burning rate. Instead, we must turn to whatever experimental evidence may be available. The point of introducing the notion of a maximum allowable burning rate has been to demonstrate that it can exert a significant influence on Diesel engine performance. Of course, the notion of "one" maximum burning rate is undoubtedly an oversimplification. More than

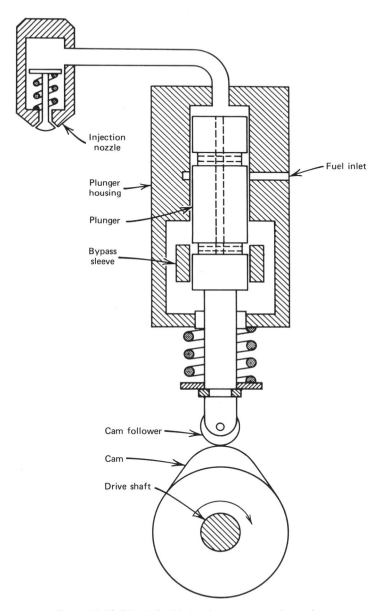

Figure 5.49 Diesel fuel injection pump and nozzle.

likely, its value changes as combustion proceeds, reflecting the decrease in numbers of available oxygen molecules. If this is true — that maximum burning rate decreases in value as combustion proceeds — then the effect on power, say, which results when the maximum burning rate is reached, will be more pronounced than is suggested in Figs. 5.46 and 5.47. Note also that the maximum burning rate is not likely to be reached in engines operating at moderate speed or in engines operating on small fuel flow, so that the duration of combustion is short to begin with.

Figure 5.49 illustrates the essential features of one design of Diesel injection pump. It is similar in some respects to the pump shown in Fig. 5.7. The differences are in the manner by which the plungers are driven: a swash plate for gasoline injection and a profiled cam for Diesel injection. The cam can be shaped to any profile we choose. Injection will begin when fuel line pressure is sufficient to open the nozzle. The quantity of fuel delivered and the manner in which it is delivered as a function of time depend on cam profile.

As a final comment, note that it is not necessary to equip Diesel engines with throttles in order to control power output. Instead, power can be controlled by fuel flow. Referring back to Fig. 5.44, point 3, the end of combustion, is fixed by the amount of fuel injected, which can be varied over a wide range of limits, down to something quite close to zero. The upper limit on fuel flow is generally set by the amount of smoke that can be tolerated in the engine exhaust plume. Presumably, the appearance of smoke suggests the presence of unburned carbon, indicating that the engine may be operating at the limit of maximum burning rate or at the limit of available oxygen or both.

5.16 HEAT TRANSFER

During one cycle of any internal combustion engine there is a net transfer of heat from the engine to the surroundings. The transfer is a matter of neccessity, for the interior surfaces of cylinder walls, cylinder heads, and piston faces must be maintained at safe operating temperatures. Heat is transferred to or from the working fluid during every part of a cycle, and the net work done by the working fluid during one complete cycle must be evaluated by stepwise numerical integration over the cycle

$$w_{\text{net}} = \oint \left(P + \frac{\Delta P}{2} \right) \Delta V$$

The pressure change, ΔP, is the result of piston motion, combustion, flow into or out of the engine, and heat transfer, the latter taking the form

$$\frac{\Delta P}{P} = \frac{h_c A (T_w - T)}{M c_v T} \Delta t \qquad\qquad 5\text{-}109$$

where h_c = heat transfer coefficient,
$\quad A$ = interior surface area of the engine volume,
$\quad T_w$ = interior surface temperature,
$\quad T$ = working fluid temperature,
$\quad M$ = mass of working fluid,
$\quad c_v$ = working fluid specific heat,

and Δt is a suitably small increment in time.

The quantities A, M, c_v, and T present no difficulties. The surface temperature T_w can be assigned, within reasonable limits, because its value depends on the design and operation of the cooling system. The problem is the choice of a proper value for the heat transfer coefficient h_c. The mode of heat transfer between the working fluid and the interior surfaces is forced convection, and the value of h_c depends, among other things, on the gas velocity at the surfaces. Since we know little about the gas motion, a large measure of uncertainty enters into any heat transfer calculations. An expression for h_c can be written

$$h_c = 0.13 \times B^{-0.2} \times P^{0.8} \times T^{-0.5} \times Z^{0.8} \ (\mathrm{kJ/m^2/s/K}) \qquad \text{5-110}$$

where

B = piston bore, m,
P = cylinder pressure, atm,
T = working fluid temperature, K,
Z = working fluid velocity, m/s.

Equation 5-110 is derived from experimental studies in forced convection, which provide a conceptual framework, and from experimental data from engine testing. Whether Eq. 5-110 is accurate within a factor of 2 or 10, or whatever, is not yet a question that can be answered definitively.

The importance of heat transfer in estimating engine performance is great. Heat transfer reduces power output. This can be seen at once by assuming T_w equal to the temperature at the end of compression. During expansion from TDC to BDC, the direction of transfer is from the working fluid, which reduces expansion work. During the exhaust stroke, transfer continues from the working fluid, which will tend to increase the quantity of residual exhaust. During the intake stroke, heat is transferred to the working fluid, which reduces the quantity of fresh charge at the start of the compression stroke, and heat transfer to the working fluid continues during the compression stroke, increasing the work of compression.

There is an abundant amount of data that show how much heat is transferred from an engine per cycle. That information is helpful, in a general sort of way, but we are not yet in a position to relate those measurements to engine geometry and operating parameters in such a way as to deal confidently with the details of heat transfer during each small increment in time of a cycle.

5.17 MAXIMUM THEORETICAL EFFICIENCIES

The thermal efficiency of an internal combustion engine is usually calculated as the ratio of the net work per unit mass of fuel burned divided by the heat of reaction, constant volume, or constant pressure, whichever is appropriate. Combustion releases a certain quantity of energy, and the thermal efficiency informs us as to what fraction has been converted into useful work. Thermal efficiency computed in that fashion is often referred to as a first law efficiency.

There is another approach to thermal efficiency, which asks the following question: When a unit mass of fuel is burned, what is the maximum fraction of the heat released that can be converted to work by a *heat engine*? Suppose unit mass of fuel is burned at constant volume, adiabatically, reaching a temperature T_f. The immediate surroundings, whether the air, the ocean, a lake, or the ground, are at temperature T_o. Using the products of combustion as a heat source and the surroundings as a heat sink, we can operate a heat engine. But the operation cannot continue indefinitely, because each cycle of operation reduces the temperature of the products. When they reach T_o, work production stops.

The extraction of maximum work during the cooling of the products of combustion requires an engine operating in a reversible cycle, very small in size, so that the amount of heat added to the engine during any cycle can be considered infinitesimal. The work delivered to the surroundings in any cycle is also infinitesimal; the total work delivered is found by summing. Thus, for one cycle

$$\delta w = \delta q_a - \delta q_r$$

δw being the work delivered, and δq_a and δq_r representing, respectively, the heat added to and rejected by the engine. Let T denote the temperature of the product gas mixture at any arbitrary time during the working life of the engine. Then, by Kelvin's theorem,

$$\frac{\delta q_a}{T} = \frac{\delta q_r}{T_o}$$

and the expression for work becomes

$$\delta w = \delta q_a \left(1 - \frac{T_o}{T} \right)$$

But

$$\delta q_a = - M c_v \, dT$$

M and c_v being the mass and heat capacity of the products. The minus sign appears because δq_a and dT have opposite signs. Eliminating δq_a and integrating between T_f and T_o

$$
\begin{aligned}
W &= -M \int_{T_f}^{T_o} c_v \left(1 - \frac{T_o}{T}\right) dT \\
&= M \left[u(T_f) - u(T_o) - T_o \int_{T_o}^{T_f} c_v \frac{dT}{T} \right] \\
&= M \left\{ u(T_f) - u(T_o) - T_o \left[\phi(T_f) - \phi(T_o) - R(T_f - T_o) \right] \right\}
\end{aligned}
\qquad \text{5-111}
$$

where
$$
u(T) = \Sigma\, x_i\, u_i(T)
$$
$$
\phi(T) = \Sigma\, x_i\, \phi_i(T)
$$

the x_i denoting mole fractions in the product gas mixture.

Some notion of the order of magnitude for the maximum possible integrated thermal efficiency can be found by assuming c_v constant and dividing Eq. 5-111 by the constant volume heat of reaction,

$$
U_{rp} = M[u(T_f) - u(T_o)]
$$

and we get

$$
\eta_{th} = 1 - \frac{\ln(T_f/T_o)}{T_f/T_o - 1} \qquad \text{(constant volume)} \qquad \text{5-112}
$$

As a numerical example, if $T_o = 300$ K and $T_f = 2900$ K, then

$$
\eta_{th} = 0.74
$$

The best we can hope to obtain from constant volume cooling of the products of combustion is 74% of the energy released by combustion. That is the upper limit set by the second law.

5.18 SUMMARY

Internal combustion engines come in a wide variety of sizes, shapes, and designs. Operating parameters are numerous and may be assigned large ranges of values. Thermodynamics provides information on thermal efficiency and theoretical or indicated power output for all possible engine/operating combinations. But certain crucial questions associated with the details of the progress of combustion pose problems that are beyond the reach of thermodynamics, even when pushed to its limit.

The combustion process and the exhaust/intake process each exert great influence on power output predictions. One of the aims of this chapter has been to demonstrate rational approaches to both processes.

The material in this chapter equips the reader with sufficient information to prepare an encyclopedic summary of the operating characteristics of any four-stroke engine.

REFERENCES

The following texts contain a wealth of information on the details of engine design and construction, as well as the results of a large body of experimental work. The emphasis is more on real engine performance and less on theory of performance, as predicted by thermodynamics. Taylor's book contains an enormous bibliography.

L. C. Lichty, *Combustion Engine Processes*, McGraw-Hill Book Co., New York, 1967.
E. F. Obert, *Internal Combustion Engines and Air Pollution*, Intext Educational Publishers, New York, 1974.
C. F. Taylor, *The Internal Combustion Engine in Theory and Practice* (2 vol.), M.I.T. Press, Cambridge, Massachusetts, 1977.

A comprehensive and lucid description of rotary engines will be found in:

R. F. Ansdale, *The Wankel RC Engine*, A. S. Barnes & Co., Cranbury, New Jersey, 1969.

The causes of and problems associated with exhaust emissions is well summarized in:

D. J. Patterson & N. A. Henein, *Emissions from Combustion Engines and Their Control*, Ann Arbor Science Publishers, Ann Arbor, Michigan, 1972.

An entertaining account of internal combustion engine history, written by a historian, is:

L. Bryant, *The Invention of the Internal Combustion Engine Technology and Culture*, Vol. VII, No. 2, 1966, and Vol. VIII, No. 2, 1967.

The published literature on theory and experiment is large but confined essentially to two journals:

Transactions of the Society of Automotive Engineers (SAE) *Combustion Science and Technology*

The latter journal is recent, begun in 1970. The SAE *Transactions* carry back to the earliest published work on engines.

PROBLEMS

5.1 Plot the thermal efficiency of the air standard Otto cycle (Eq. 5-8) against compression ratio, for $k = 1.4$

5.2 For octane, C_8H_{18}, plot the temperature drop resulting from evaporation (Eq. 5-7) against fuel/air equivalence (fuel/air ratio divided by chemically correct fuel/air ratio).

5.3 Repeat Problem 5.2 for methyl alcohol (methanol), CH_3OH, and for ethyl alcohol (ethanol), C_2H_5OH.

5.4 Write a computer program for an Otto cycle engine equipped with a carburetor and fueled with C_8H_{18}. Assume instantaneous combustion and no supercharging, and model the exhaust/intake process according to Section 5.6, so that $T1$ is obtained from Eq. 5-44, which is the general case.

The program may be used to show how net work, or power at some specified speed, varies with

Compression ratio
Intake manifold temperature
Intake manifold pressure
Exhaust manifold pressure
Fuel/air ratio

The program may be checked against Tables 5.2, 5.3, 5.4, and 5.5.

5.5 Fuels that can be derived from organic matter have been suggested as substitutes in combustion engines for petroleum-based fuels. Methyl alcohol is an example.

Alter the program in Problem 5.4 to examine the performance of an engine fueled with a CH_3OH/C_8H_{18} mixture. Vary the mixture from pure CH_3OH to almost pure C_8H_{18}.

In view of the large temperature drop that accompanies evaporation of methyl alcohol (see Problem 5.3), severe icing will probably occur when gasoline is diluted with alcohol. This undesirable, and possibly dangerous, condition can be offset by transferring heat to the intake manifold from the exhaust system. Make the calculations, assuming that intake manifold temperature $TM = 300$ K for all fuel combinations and for all fuel/air ratios.

5.6 Liquid hydrogen is a possible combustion engine fuel, provided that energy from solar collectors or from a hydroelectric installation is available to produce the hydrogen. Storage is a problem with liquid hydrogen. On the other hand, the products of combustion will not contain CO or unburned hydrocarbons. The exhaust, however, will contain NO_x.

Alter the program in Problem 5.4 to accept hydrogen as the fuel. Maintain the intake manifold temperature constant at 300 K, or some other reasonable level, assuming heat transfer from the exhaust system. Since hydrogen/air mixtures do not detonate, hydrogen-fueled engines can probably be operated at considerably higher compression ratios, which favors higher thermal efficiency.

5.7   Power output from Otto cycle motors is controlled by the throttle at the entrance of the intake manifold. When throttled, an engine works against itself, and fuel is wasted overcoming work done during the exhaust/intake loop.

A proposal to avoid that waste is accomplished by deactivating half the cylinders; the rocker arms are made inoperative, so intake and exhaust valves in the deactivated cylinders remain closed at all times. The net effect is to reduce the motor displacement volume to half its full value. (When an engine operates with N, rather than $2N$ cylinders, the torque may become unacceptably rough. That problem will not concern us here.)

Suppose an engine is operating at some specified speed, producing a certain amount of power, and running at part throttle. Now deactivate half the displacement volume. In order to produce the same power, at the original speed and fuel/air ratio, the intake manifold pressure must be increased. The engine therefore moves to a new condition of operation.

What savings in fuel consumption are possible?

5.8   Modify the computer program of Problem 5.4 to accommodate a gear-driven supercharger. For the same speed and mixture ratio, power can be increased.

Superchargers are particularly useful in aircraft engines, since they compensate for the fall in ambient pressure with altitude. Suppose an engine is equipped with a compressor capable of producing a 1.5 pressure ratio between inlet and outlet and operates at 90% efficiency. To what altitude can sea level power at full throttle be maintained? Assume "standard" atmosphere, that is:

Altitude (feet)	$t(°F)$	P/P_o
0	59	1.000
5,000	41	0.832
10,000	23	0.688
15,000	6	0.564
20,000	−12	0.460

The computations may be made in several ways. We could assume that temperature in the intake manifold remains constant, by heating, at, say, 300 K. Or, ignoring the possibility of icing, or rather the certainty of icing, let air temperature entering the carburetor be ambient and take into account the temperature drop resulting from fuel evaporation.

The scavenging process may be treated in two ways. The usual calculation assumes all exhaust gas to be swept out at top dead center, owing to the pressure difference between intake manifold and exhaust manifold, the latter taken to be ambient pressure. With this assumption, the cycle can be completed in a single pass. On the other hand, when residual exhaust is trapped in the cylinder at top dead center, the intake stroke must be reexamined because then Eq. 5-44 is no longer applicable. As the intake process begins, fresh charge compresses the residual exhaust to intake manifold pressure, after which the piston moves toward bottom dead center, drawing in additional fresh charge. Analysis leads to a new form for Eq. 5-44.

5.9 Rewrite the computer program of Problem 5.4 to include an exhaust turbine supercharger through which all the exhaust passes. Select values for compressor and turbine efficiencies and for the intake manifold temperature, and then determine the altitude to which full throttle sea level power can be maintained.

5.10 Verify Eqs. 5-4 and 5-5.

5.11 Modify the program of Problem 5.4 to include progressive burning. The exhaust/intake process retains the simple form discussed in Section 5.6. Now the following parameters, in addition to those already listed in Problem 5.4, can be explored to discover their influence on motor operation:

Speed
Spark advance
Duration of combustion
Rate of combustion

The program can be checked against Fig. 5.24a and b.

5.12 Using the program developed in Problem 5.11, reexamine the following problems:

Problem 5.7	Dual displacement volume engine
Problem 5.8	Altitude performance with gear-driven supercharger
Problem 5.9	Altitude performance with exhaust supercharger

5.13 Using the program of Problem 5.11, Problems 5.5 and 5.6 can be reexamined. However, the results of such investigations cannot be regarded as either useful or predictive until we know something about the burning characteristics of hydrogen or CH_3OH/C_8H_{18} mixtures with air. The reason, of course, is the central position of importance occupied by $\Delta\theta_c$, the duration of combustion, as shown in Fig. 5.24a. Spark advance may be set arbitrarily. Depending on combustion chamber goemetry, Eq. 5-70, 5-71, or 5-72 could be appropriate. But we should be prepared to discover that the speeds of burning fronts in hydrogen/air mixtures and C_8H_{18}/air mixtures may be quite different, and flame speed exerts a controlling influence on the time required for combustion. Consequently, to reexamine Problems 5.5 and 5.6, it is first necessary to search combustion literature with the hope of finding experimental results that can lend some support to conjectures concerning $\Delta\theta_c$ for exotic engine fuels.

5.14 Modify the program of Problem 5.11 to include analysis of the exhaust and intake strokes, as described in Section 5.11. The following parameters, in addition to those listed in Problems 5.4 and 5.11, can now be altered:

Wide-open effective area of exhaust valve
Wide-open effective area of intake valve
Exhaust valve area as function of crank angle
Intake valve area as function of crank angle

The program may be checked against Tables 5.8 through 5.14 and Figs. 5.37 and 5.38.

Structural considerations, namely, the strength of the cylinder head, control maximum allowable valve dimensions. The dynamics of valve, rocker arm, push rod, and cam follower motions place practical limits on maximum valve lift. Together, valve lift and valve diameter fix the wide-open valve area, which must then be modified by a suitable flow coefficient, which in turn depends on valve geometry.

In the text, wide-open area AO and lift curves were equal for both valves for simplicity in reporting results. In practice, the exhaust valve diameter is usually larger than that of the intake valve, whereas maximum lift is about identical for both valves.

5.15   Using the program of Problem 5.14, the following problems may be reexamined:

Problem 5.7	Dual displacement volume engine
Problem 5.8	Altitude performance with gear-driven supercharger
Problem 5.9	Altitude performance with exhaust turbine supercharger

The effects of various valve timings on power output variation with altitude are worth examining.

5.16   Problems 5.5 and 5.6 can be reexamined with the program of Problem 5.14. The comments in Problem 5.13 apply; that is, some comparative figure related to the burning rates of various fuels should be found in order to make meaningful any comparison of engine performance with those fuels.

5.17   Table 5.12 shows, among other things, that at wide-open throttle, a specified power can be obtained at more than one speed by altering exhaust valve opening and intake valve closing. That observation suggests the following: for some specified power, what combination of valve timing and speed produce the maximum thermal efficiency?

5.18   A gear-driven supercharger is clearly worthless if the increase in work done by the working fluid during the engine cycle is offset by the work required to drive the compressor. For a series of selected pressure ratios in the range, say, of 1.1 to 1.6, and for some selected speed, what are the associated minimum acceptable compressor efficiencies?

5.19   A fuel is burned and the products subsequently cooled to ambient temperature, with pressure remaining constant during the cooling process. Derive the counterparts to Eqs. 5-111 and 5-112.

5.20   Plot the thermal efficiency of the air standard Diesel cycle (Eq. 5-96) against compression ratio, for $k = 1.4$ and for volume ratios $VR = 1$, 2, and 4. Compare with the plot of Problem 5.1.

5.21   Write a computer program for a Diesel cycle engine, using $C_{10}H_{22}$ fuel. Assume constant pressure combustion starting at top dead center, and model the exhaust/intake processes according to Section 5.6. The program will reveal how work output for a cycle is affected by:

Compression ratio
Fuel/air ratio
Air intake temperature
Air intake pressure
Exhaust pressure

The program can be checked against Table 5.16.

5.22   As with an Otto engine, the ideal Diesel engine can be equipped with a gear-driven supercharger (Problem 5.8) or an exhaust turbine-driven supercharger (Problem 5.9).

5.23   Modify the program of Problem 5.21 to incorporate the following:
(*a*) The quantity of fuel injected during each increment in crankshaft advance may be specified arbitrarily. Equation 5-101 is the special case for constant pressure combustion. Any functional relationship between n and V is acceptable, because the cam profile in Fig. 5.49 can be shaped in any way we please.
(*b*) Fuel injection may be initiated at any point in the cycle. Equation 5-108 is an example, when Eq. 5-101 relates n and V. According to the example illustrated in Fig. 5.48, injection timing can have a marked impact on power.
(*c*) A maximum burning rate $(dn/dt)_{max}$ is imposed. According to the example in Fig. 5.47, maximum burning rate is not as influential as injection timing. Unfortunately, thermodynamics is silent on the matter of a quantitative measure for the maximum burning rate, and we must resort to whatever information can be squeezed out of experimental observation. The important notion is that there is a maximum burning rate; our intuition so informs us. The analysis reveals that constant pressure combustion cannot be maintained when that burning rate limit is reached.

5.24   Modify the program of Problem 5.23 to include the analysis of intake and exhaust strokes as discussed and elaborated in Section 5.11. The remarks included in Problem 5.14 for the Otto cycle engine can be applied here to the Diesel cycle engine. The difference in programming between this problem and 5.14 is reduced to a single item: in the Otto cycle the intake manifold contains (for a carbureted engine) a mixture of air and fuel vapor, whereas the intake manifold of a Diesel contains only air.

6

TWO-STROKE INTERNAL COMBUSTION ENGINES

Thermodynamic analysis of two-stroke engines leads to performance predictions that are more open to question than those obtained for four-stroke engines. The uncertainty lies in the scavenging process of the two-stroke cycle, which is the subject of this chapter. The other processes — compression, combustion, and expansion — may be treated with the formulations developed in Chapter 5.

Figure 6.1 illustrates the two-stroke engine that is to be analyzed. In place of mechanically actuated valves, the flow of gases into and out of the cylinder occurs through ports machined in the cylinder wall. Generally, the ports are diametrically opposite each other and are either open or closed, depending on piston position. This simple porting arrangement permits two-stroke engines to operate at higher speeds than four-stroke engines with comparable bore and stroke.

The engine is equipped with a transfer passage between crankcase and cylinder and with two check valves. On the upward piston stroke, the transfer passage check valve will close, and fuel/air mixture in Otto engines or air in Diesel engines is drawn into the crankcase. On the downward piston stroke, the crankcase check valve closes, and gas in the crankcase is compressed and subsequently flows into the cylinder when the intake port is uncovered and cylinder pressure drops below crankcase pressure. As drawn in the figure, the ports open and close simultaneously. In many engines the opening of the exhaust port occurs first on the downward piston stroke and last on the upward stroke. This detail does not alter the analysis in what follows.

An ideal pressure-volume cycle for a Otto engine is drawn in Fig. 6.2. A plot of crankcase pressure is included. Starting at point 1, with both ports closed, the cycle passes through compression, combustion, and expansion, arriving at point 4. Now the ports are uncovered, cylinder pressure drops to crankcase pressure, crankcase gas passes into the cylinder through the transfer passage, mixing occurs in the cylinder,

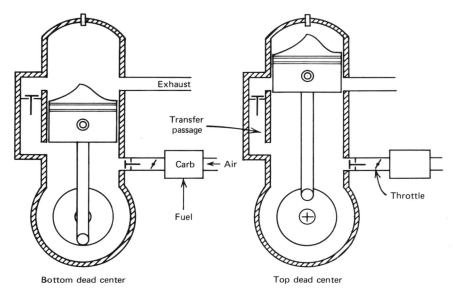

Figure 6.1 A two-stroke engine.

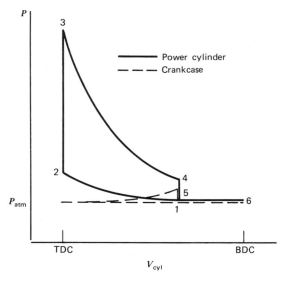

Figure 6.2 Ideal two-stroke Otto cycle, full throttle.

and some gases leave the cylinder. This complicated set of events in the cycle ends at point 5, when cylinder and crankcase pressures are both atmospheric. With further piston movement, more gas transfers from crankcase to cylinder until BDC is reached at point 6. On the return to point 1, the crankcase check valve opens, the transfer passage check valve closes, and a portion of the gases in the cylinder is displaced into the exhaust port and out of the engine.

Figure 6.3 illustrates the engine at point 4, the beginning of the scavenging and mixing process just described. How shall we reduce these processes to a set of events that can be analyzed? In particular, how shall we deal with the mixing process as gas passes from the crankcase into the cylinder, points 4 to 5? There are three possible lines of analysis.

1. We can assume that gas entering the cylinder from the transfer passage "pushes" the products of combustion ahead until the level of pressure has dropped to atmospheric. The gases in the cylinder then mix and come to a uniform temperature. Ideally, none of the fuel/air mixture entering the cylinder would have passed out of the cylinder; only combustion products leave. This assumption leads to minimum residual exhaust gas and could be considered for engines with a very large bore/stroke ratio.
2. We can assume that fuel/air mixture from the transfer passage "bores" through the combustion products to the exhaust port opposite, leaving the bulk of product gases largely undisturbed. In this case, only a small amount

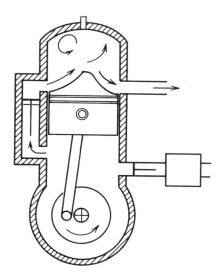

Figure 6.3 Piston position at the start of the scavenging process, point 4.

of exhaust is scavenged, as might happen in engines with a very large stroke/bore ratio.
3. We can assume that fuel/air mixture and combustion products mix continuously, and a mixture of both leaves the cylinder between points 4 and 5. With this model, scavenging efficiency can be expected to fall somewhere between the extremes represented by the first two assumptions.

We shall examine the third possibility, since it is the moderate compromise. As shown in Fig. 6.3, the piston face is contoured in order to produce a swirling motion in the fuel/air mixture, thus deflecting its movement away from the exhaust port and promoting maximum mixing. We then have a mixing problem, shown in Fig. 6.4 and described as follows:

A fixed volume contains $X1$ moles of combustion products at temperature TX. (From the cycle calculations, both are known.) Into this volume will flow $M1$ moles of fuel/air mixture which is at TM. (Both of these are known; $M1$ is the fuel/air mixture drawn into the crankcase on that portion of the upward piston stroke between points 1 and 2. The air inlet temperature at the carburetor and the fuel/air ratio fix TM.) There is no heat transfer to the surroundings, and the pressure is everywhere constant. The problem is to determine the amount of fuel/air mixture, $M2$, and the amount of residual exhaust, $X2$, in the cylinder at the conclusion of mixing, when the temperature is $T5$.

This problem is a considerable simplification over real events taking place in the engine between points 4 and 5, Fig. 6.2. During downward piston stroke, fuel/air mixture in the crankcase is compressed, which raises its temperature above TM. And if pressure were everywhere constant, neither gas, fuel/air mixture, nor combustion products would move. However, with a large crankcase volume relative to piston

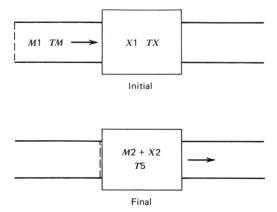

Figure 6.4 A mixing problem.

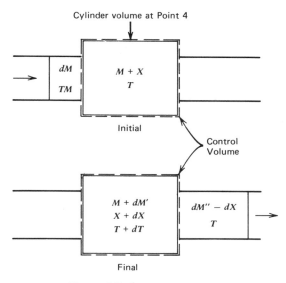

Figure 6.5 A mixing problem.

displacement volume, the compression is not appreciable, and the approximations become reasonable.

To analyze the mixing process, examine the state of the cylinder at some time t, as illustrated in Fig. 6.5. The cylinder contains M moles of fuel/air mixture and X moles of combustion products at temperature T. In a brief time interval, dM moles of fuel/air mixture enter at temperature TM. The mole numbers change to $M + dM'$ and $X + dX$ and the temperature becomes $T + dT$. Simultaneously, dM'' moles of fuel/air mixture leave the cylinder, along with $-dX$ moles of combustion gases. (Conservation of matter calls for the minus sign. Clearly, dX is a negative quantity, because X decreases with time.) To analyze the problem we make use of four statements:

1. Conservation of fuel/air mixture
2. Mixing is complete
3. Pressure and volume remain constant
4. Energy is conserved

In addition, the gases are ideal, and matters are further simplified by ignoring differences in heat capacities. Taking these statements in the order listed, conservation of matter for the fuel/air mixture requires

$$dM = dM' + dM''$$

Complete mixing means that the mixture composition in the cylinder is everywhere uniform. Stated mathematically

$$\frac{M + dM'}{X + dX} = \frac{dM''}{-dX}$$

which states that the proportions of fuel/air mixture and combustion products in the cylinder are identical to those in the mixture that leave the cylinder. But this equation may be reduced to

$$\frac{M}{X} = -\frac{dM''}{dX}$$

by cross-multiplying and dropping out the second-order terms. Combining these two statements, we get

$$dM' = dM + \frac{M}{X} dX \qquad\qquad 6\text{-}1$$

$$dM'' = -\frac{M}{X} dX \qquad\qquad 6\text{-}2$$

Since pressure and temperature remain constant

$$(M + X)\,T = X1TX = \text{constant}$$

Differentiating and noting that the differential of M is dM',

$$\frac{dM' + dX}{M + X} = -\frac{dT}{T}$$

and eliminating dM' with Eq. 6-1

$$\frac{dM}{M + X} + \frac{dX}{X} = -\frac{dT}{T} \qquad\qquad 6\text{-}3$$

To construct the mathematical statement of energy conservation, start with the initial state at time t. The energy in the control volume is

$$c_v(M + X)T \qquad\qquad (a)$$

During time dt the energy entering the control volume is

$$c_p\,TM\,dM \qquad\qquad (b)$$

and the energy leaving is

$$c_p T(dM'' - dX) \tag{c}$$

which can be rewritten as

$$- c_p T \frac{M + X}{X} dX \tag{d}$$

by eliminating dM'' with Eq. 6-2. In the final state the energy in the control volume is

$$c_v(M + X + dM' + dX)(T + dT) \tag{e}$$

Eliminating dM', the line reads

$$c_v \left(M + X + dM + \frac{M + X}{X} dX \right)(T + dT) \tag{f}$$

and making use of Eq. 6-3, the line reduces to

$$c_v(M + X)T \tag{g}$$

which is identical to the initial energy in the control volume, which results from the requirement that pressure and volume remain constant. Consequently, conservation of energy reduces to an equality between entering and leaving energy quantities, or

$$TM\, dM = - T \frac{M + X}{X} dX \tag{6-4}$$

But $(M + X)T = X1\,TX$, and by substitution, the energy equation changes to

$$TM\, dM = - X1\ TX \frac{dX}{X}$$

which can be integrated

$$TM \int_0^{M1} dM = - X1\ TX \int_{X1}^{X2} \frac{dX}{X}$$

and the final result is

$$X2 = X1\ e^E, \qquad \text{where} \qquad E = - \frac{M1\ TM}{X1\ TX} \tag{6-5}$$

which is physically acceptable. For we expect $X2$ to be small when $M1$ is large; Eq. 6-5 puts the matter precisely and informs us as to just how $X2$ depends on $M1$, along with $X1$, TM, and TX.

To discover the final temperature, $T5$, eliminate dM between Eqs. 6-3 and 6-4, and the result is

$$\frac{TM}{T(T-TM)}dT = \frac{dX}{X}$$

which integrates to

$$T5 = \frac{TM}{1 - \frac{TX - TM}{TX}e^E} \qquad\qquad 6\text{-}6$$

Using Eqs. 6-5 and 6-6, we can now follow the entire scavenging process as the events move from points 4 to 5 to 6 to 1. Let PA denote atmospheric pressure. Then $X1$, which denotes the number of moles of combustion products in the cylinder when mixing with fuel/air mixture commences, is

$$X1 = \frac{PA\,V1}{R\,TX}, \qquad TX = T4\left(\frac{PA}{P4}\right)^{(k_p-1)/k_p} \qquad\qquad 6\text{-}7$$

That formulation for $X1$ is the smallest possible value we can assign. Small $X1$ leads to large E in Eq. 6-5 and thus to small $X2$. Consequently, Eq. 6-7 puts matters on the optimistic side. By definition, Fig. 6.4, $M1$ denotes the moles of fuel/air mixture that will flow into the cylinder. After $M1$ has moved from crankcase to cylinder, crankcase pressure is atmospheric. Consequently, $M1$ can be equated to the amount of fuel/air mixture that was drawn into the crankcase during upward piston motion between points 1 and 2,

$$M1 = \frac{PA(V1-V2)}{R\,TM} \qquad\qquad 6\text{-}8$$

For the quantity E occurring in Eq. 6-5,

$$E = -\frac{M1\,TM}{X1\,TX} = -\frac{V1-V2}{V1} = \frac{1}{CR} - 1 \qquad\qquad 6\text{-}9$$

where

$$CR = \text{compression ratio} = V1/V2. \qquad\qquad 6\text{-}10$$

Let $NX5$ denote the moles of combustion products in the cylinder at the conclusion of the mixing process. Then $NX5$ will be given by the equation for $X2$, Eq. 6-5, or

$$NX5 = \frac{PA \; V1}{R \; T5} \, e^E$$

and by subtraction, $NM5$, the moles of fuel/air mixture in the cylinder at the conclusion of the mixing process, is

$$NM5 = \frac{PA \; V1}{R \; T5} - NX5$$

The piston now moves slowly from point 5 to 6, and in so doing

$$\frac{PA \; (V6 - V1)}{R \; TM}$$

moles of fuel/air mixture will be transferred from the crankcase to the cylinder, where mixing with

$$NX5 + NM5$$

will occur, producing a mixture temperature given by

$$T6 = \frac{(NM5 + NX5) \; T5 + \dfrac{PA(V6 - V1)}{R \; TM} \; TM}{(NM5 + NX5) + \dfrac{PA(V6 - V1)}{R \; TM}} \qquad \text{6-11}$$

During that mixing, which takes place at constant pressure, nothing leaves the cylinder. But on the return piston stroke, from points 6 to 1, a portion of the mixture of combustion gases and fuel/air mixture will be pushed out of the cylinder (the transfer passage check valve will close as the piston moves upward). Thus the moles of residual exhaust gas, $NX1$, and fuel/air mixture, $NM1$, which will be found in the cylinder at the start of the next cycle calculation, are given by

$$NX1 = \frac{NX5}{SR}$$

$$NM1 = \left[NM5 + \frac{PA \; (V6 - V1)}{R \; TM} \right] \frac{1}{SR} \qquad \text{6-12}$$

where SR = scavenging ratio = $V6/V1$, and $T1 = T6$, from Eq. 6-11.   6-13

The procedure now is to perform a sufficient number of cycle calculations to achieve convergence; that is, the numbers repeat themselves between successive cycles. (For low compression ratios in the region of 5:1, this may require four cycles.) Finally, the net work output is computed from

$$WNET = WEXP - WCOMP - WCC$$

where WEXP = work done by the exhaust gases during isentropic expansion between
              point 3 and 4
          $= U_p(T3) - U_p(T4)$
WCOMP = work done on the fuel/air mixture and residual exhaust during isen-
              tropic compression between points 1 and 2
          $= (NM1 + NX1)c_v(T2 - T1)$
   WCC = net work done on the fuel/air mixture in the crankcase during a com-
              plete cycle.

To evaluate WCC, note first that the quantity of fuel/air mixture drawn into the crankcase during the complete upward stroke of the piston, from point 6 to point 1, is

$$\frac{PA(V6 - V2)}{R\,TM}$$

Now let VC represent the volume in the crankcase with piston as TDC, and further introduce

$$VR = \frac{VC}{V1 - V2}$$ 6-14

The work done on the gases in the crankcase during a complete cycle can be written

$$WCC = -\int_{V2}^{V1} P\,dV - PA(V1 - V2)$$

where P and V are crankcase pressure and volume. Assuming the compression from point 2 to i is isentropic

$$WCC = PA\,V2\,(CR - 1)\left[\frac{VR}{k_r - 1}\left\{\left(\frac{VR}{VR - 1}\right)^{k_r - 1} - 1\right\} - 1\right]$$ 6-15

where k_r is the specific heat ratio for the fuel/air mixture in Otto engines, or, for air in Diesel and fuel-injected Otto engines.

This completes the full throttle analysis. The combustion process has not been discussed, for it may be treated by the methods for Otto and Diesel engines already described in Chapter 5. As indicated in Fig. 5.39, rate of burning, or total combustion time, is affected by the fraction of residual exhaust gas in the mixture at ignition. As will be shown shortly, two-stroke engines regularly operate with large residual exhaust fraction, which suggests that combustion time in two-stroke engines, as compared to four-stroke engines, may occupy a larger span of crankshaft rotation.

It is worth noting that two-stroke engines can be characterized by two thermal efficiencies. One is the customary thermal efficiency of the cycle,

$$\eta_{th}. \text{ cycle} = \frac{\text{net work of one cycle}}{\text{fuel burned in one cycle} \times (-U_{rp})}$$

while the other is based on the amount of fuel pumped into the engine, the engine efficiency,

$$\eta_{th}, \text{ engine} = \frac{\text{net work of one cycle}}{\text{fuel pumped in one cycle} \times (-U_{rp})}$$

which would apply to spark ignition Otto engines with carburetors. Because of the manner in which the scavenging process was treated, some of the fuel/air mixture will pass through the engine and never burn. Engine efficiency will always be less than cycle efficiency. For fuel-injected Otto engines and for Diesel engines, the two efficiencies can be identical.

Turning the discussion to some performance computation results, a comprehensive survey of two-stroke engine operation can become extensive and involved. Whereas four-stroke engines are characterized by only two volumes, four are required to specify the kinematics of a two-stroke engine of the type shown earlier in Fig. 6.1. Referring to Fig. 6.6, any three of the four volumes may be assigned and the fourth varied over a limited but nevertheless wide range. For example, fixing $V2$, $V6$, and VC assigns piston positions at TDC and BDC and specifies the crankcase volume with the piston at TDC. Volume $V1$, the start of compression, can be varied by altering the locations of the intake and exhaust ports in the cylinder wall, with attendant changes in the ratios CR, VR, and SR. Figure 6.7 illustrates how the two thermal efficiencies of a carbureted Otto engine will respond. Included in that figure is the ratio of the two thermal efficiencies, which is simply the ratio of fuel burned to fuel pumped. Note, for example, that a 6:1 compression ratio engine will, for full-throttle conditions, burn only half the fuel pumped through the carburetor. As with four-stroke engines, thermal efficiency, by either measure, improves with increasing compression ratio, with a tendency toward leveling off with high compression ratio.

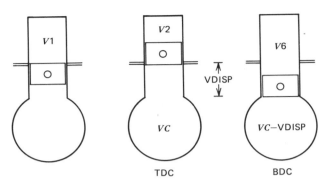

Figure 6.6 Two stroke engine volume relations. CR = compression ratio = $V1/V2$; SR = scavenging ratio = $V6/V1$; VR = crankcase ratio = $VC/(V1-V2)$; VDISP = displacement volume = $V6-V2 = V2(CR\ SR - 1)$.

On the other hand, suppose $V1$, $V2$, and VC are fixed. Piston positions at TDC and start of compression are then assigned, which fixes the compression ratio and VR. The piston position at BDC can now vary, and the scavenging ratio SR will change. Figure 6.8 shows some results. Included in the plot is the ratio $HP/V1$, power per unit of cylinder volume, measured at the start of compression. With increasing SR the cycle thermal efficiency is virtually unchanged through changes in VR, but the engine efficiency deteriorates badly, while power output improves. At large SR, a large fraction of the fuel pumped never burns, but most of the exhaust products are scavenged, which accounts for the gain in power. Other combinations of volume variation can be explored. They are left to readers with an interest in two-stroke engines.

To continue our examination of the scavenging process, consider an engine as in Fig. 6.1 for which bore and stroke have been specified. Then $V6 - V2$ is known. Suppose $VC = 2(V6 - V2)$. If we specify the compression ratio CR, we can then vary SR. Table 6.1 shows some computed results.

As SR increases, power output falls off, as does engine thermal efficiency. The table indicates that for best power for the given bore, stroke, and compression ratio, we should have $SR = V6/V1 = 1$, which is a practical impossibility, for it means the exhaust and intake ports are not uncovered until the piston reaches BDC. The entire scavenging process must take place instantly.

A dynamic analysis, taking piston rate of travel into consideration, is more complicated in a two-stroke engine of the sort shown in Fig. 6.1 than in a four-stroke engine, as discussed in Section 5.11. Figure 6.9 illustrates another way of diagramming the events that go on in the working space and crankcase of the engine. The whole exhaust/intake process is carried out in the arc extending from point 4 to 1, a fraction of a full revolution, and the time available decreases steadily as engine speed increases.

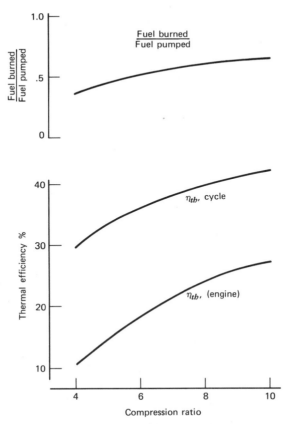

Figure 6.7 Ideal two-stroke Otto engine (Fig. 6.1). Chemically correct C_8H_{18}/Air mixture. Full throttle, TM = 300 K, $V2$ = 20, $V6$ = 220, VC = 600 (arbitrary units). Compression ratio varied by moving the location of intake and exhaust ports (see Fig. 6.6).

As engine speed increases, we can expect point 5 in the diagram to drift toward point 6, because point 5 is identified as piston position when cylinder and crankcase pressure are equal. As point 5 moves toward point 6, the time available for fuel/air transfer from crankcase to cylinder will decrease. At point 6 we wish to have that transfer completed, so that pressure at point 6 is atmospheric.

A relationship between scavenging ratio SR and $\theta 4$, crank angle at point 4, is easily developed. By definition,

$$SR = \frac{V6}{V1} = \frac{V6}{V4}$$

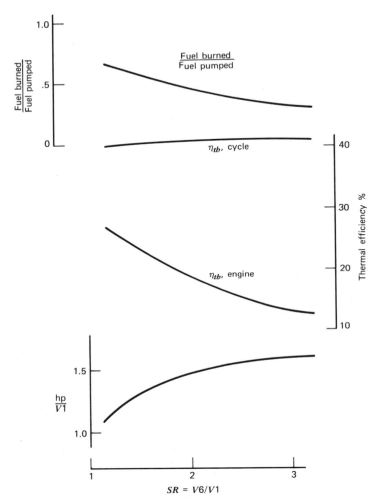

Figure 6.8 Ideal two-stroke Otto engine (Fig. 6.1). Chemically correct C_8H_{18}/Air mixture. Full throttle, **TM** = 300 K, **CR** = 8, **VR** = 4.3 (**V**1, **V**2, and **VC** constant). **V**6, piston position at BDC, varied (see Fig. 6.6).

From Eq. 5-4 the following relationship for a reciprocating piston engine is a good approximation near BDC,

$$V4 = \text{VBDC} - \text{VDISP} \; \frac{1 - \cos\theta 4}{2}$$

TABLE 6.1
BORE = 10.16 cm; STROKE = BORE; CON ROD LENGTH = 2 x BORE
VDISP = 823 cc, VC = 2 x VDISP; CR = 6; 3000 rpm; FULL THROTTLE; CHEMICALLY CORRECT
C_8H_{18}/AIR MIXTURE; TM = 300 K; COMBUSTION TIME = 60°; SPARK ADVANCE = 20°;
$n = (t/t_c)^2$.

SR	$V1$ (cc)	$V2$ (cc) TDC	$V6$ (cc) BDC	VR	Power		NM	Thermal Efficiency	
					hp	kW	$NX + NM$	Engine	Cycle
1.0	988	165	988	2.0	29.1	21.7	0.57	15.4	31.6
1.2	797	133	957	2.5	28.8	21.5	0.66	15.3	31.6
1.4	668	111	935	3.0	27.1	20.2	0.72	14.4	31.3
1.6	575	96	919	3.4	25.0	18.7	0.77	13.3	30.7

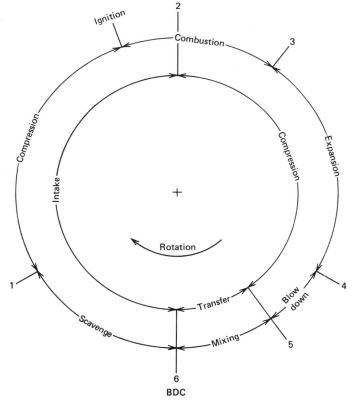

Figure 6.9 Two-stroke Otto cycle engine. Inner circle describes events in the crank-case. Outer circle describes events in the cylinder. Numbers refer to points on the pressure-volume diagram, Fig. 6.2.

which can be solved for $\theta 4$

$$\theta 4 = \cos^{-1}\left[1 - \frac{2\,CR\,(SR-1)}{SR\,CR - 1}\right]$$ 6-16

which is plotted out in Fig. 6.10. For any engine speed, that curve facilitates computation of the available time between the instant the exhaust valve cracks open and the time the piston reaches BDC.

When piston speed enters the analysis, the flow equations 5-85, 5-86, and 5-87 provide expressions for computing rates of gas flow through exhaust and intake ports. As the piston moves, after uncovering a port, the available port area for flow increases with further piston movement. The essentials with which to tackle a dynamic analysis are therefore available. With the engine reduced to the principal elements that must be considered, as in Fig. 6.11, and following a pattern similar in some details to the material in Section 5.11, the exhaust and intake processes can be examined in a time frame. The analysis is not as simple as the systems discussed in the four-stroke engine, Fig. 5.33a and b, for in Fig. 6.11 the flow patterns are more complex, particularly during the period when there are flows into and out of the cylinder. The results should, of course, be more realistic than predictions based on the model of a piston moving at infinitely slow speed.

There are several obvious modifications that can be incorporated in a two-stroke engine of the kind illustrated in Fig. 6.1 that will lead to improved thermal efficiencies:

1. Mechanically operated intake valve
2. Direct cylinder head injection
3. Supercharging

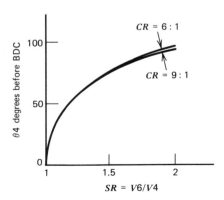

Figure 6.10 Plot of Eq. 6-16.

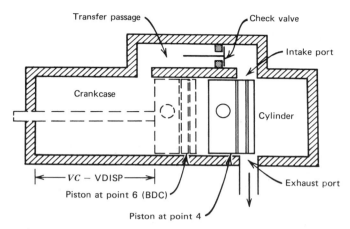

Figure 6.11 Details of the exhaust and mixing processes in a two-stroke engine.

With mechanical actuation of the intake valve, that component can be moved to the top of the cylinder, thus providing a better flow pattern of fuel/air mixture or air from the intake pipe. Such a modification will not, however, alter the thermodynamic analysis. On the other hand, with direct cylinder head injection, fuel can be injected during the most advantageous period in the Otto cycle, thereby improving thermal efficiency of the engine. Presumably, thermal efficiency of the cycle would not be changed. Supercharging can improve cycle and engine thermal efficiencies and can increase power output by reducing the amount of residual exhaust gas carried into the next cycle of operation. These modifications are attractive possibilities for two-stroke engines.

PROBLEMS

An obvious and important question is the theoretical comparison between two- and four-stroke engine performance. These first two problems are directed to that question.

6.1   Write a computer program for two-stroke Otto cycle operation. Since progressive combustion is a far more realistic assumption than instantaneous combustion, assume the former. This program is then the counterpart to the program of Problem 5.11, the four-stroke Otto engine.

Now compare the efficiencies and power for two engines with the same displacement volume and compression ratio. In the absence of precise information on the variation of duration of combustion with residual exhaust content; the combustion duration must be assumed to be equal for the engines.

Displacement volume and compression ratio completely specify the geometry of four-stroke, but not two-stroke, engines (see Fig. 6.6.) Consequently, it may be possible to assign some features of the two-stroke engine to advantage.

6.2   Problem 6.1 can be rewritten for Diesel engines. Here the combustion would be assumed to start at TDC and to proceed at constant pressure. It could also be assumed, for simplicity, that maximum burning rate, whatever it might be, is not encountered.

6.3   This problem is rather academic, albeit interesting. It assumes that an engine can be operated either as spark ignition or as compression ignition.

Diesel operation should be more efficient than Otto operation for any two-stroke engine. This does not answer the question: For the same compression ratio, how do the power outputs of the two engines compare? This question is easily disposed of with the programs in the two previous problems.

6.4   An important problem, alluded to only briefly in the text in connection with Figs. 6.9 and 6.11, is analysis of the exhaust and intake processes, taking into account the rate of piston movement with time. In the case of four-stroke engine operation, discussed in Chapter 5, there is a speed beyond which power output no longer increases but instead decreases. It seems altogether likely that two-stroke engines will exhibit approximately the same sort of behavior. Consequently, speed is an important parameter in two-stroke engine analysis. But, as pointed out in this chapter, dynamic analysis of gas flow into and out of a two-stroke engine cylinder is difficult. Nevertheless, the reader is encouraged to examine the problem.

7
ROCKETS

This chapter is concerned with the performance calculations of rockets powered by the combustion of liquid propellants. (Solid propellants are discussed in Chapter 15).

7.1 BASIC ROCKET EQUATIONS

The essential characteristic of a rocket can be understood most easily by examining motion at some remote region in space where forces resulting from atmospheric drag and gravitational attraction can be neglected. In Fig. 7.1 at some time t the rocket has mass M and is moving with velocity V.

At a later time, the mass and velocity will be $M + \Delta M$ and $V + \Delta V$. (Follow the rules of the calculus.) By conservation of matter, the ejected mass must be $-\Delta M$. (Clearly, ΔM is negative; i.e., M decreases with time.) Let Ve denote the velocity of the combustion products *relative to the rocket*. No external forces are present, so there is no change in momentum, and

$$MV = (M + \Delta M)(V + \Delta V) + (V - Ve)(-\Delta M)$$

When second-order terms are dropped, the fundamental equation for rocket motion is obtained

$$M \, \Delta V = -Ve \, \Delta M \qquad\qquad 7\text{-}1$$

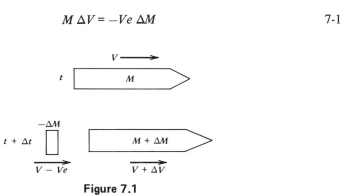

Figure 7.1

For liquid propellants, Ve is usually constant. If the rocket mass is Mo at velocity Vo, then Eq. 7-1 may be integrated to give

$$V - Vo = Ve \ln \left(\frac{Mo}{M} \right) \qquad 7\text{-}2$$

According to Eq. 7-2, when the rocket mass changes (i.e., decreases) from Mo to M, the velocity will change (i.e., increase) from Vo to V. The amount of velocity change is directly proportional to Ve. Fig. 7.2 is a plot of Eq. 7-2. Note that Eq. 7-2 does not contain time. In remote space, the velocity gain is independent of the rate at which propellants are burned.

Returning to Eq. 7-1, divide through by the time interval dt. Then

$$M \frac{dV}{dt} = Ve \left(-\frac{dM}{dt} \right) \qquad 7\text{-}3$$

The left-hand side is the product of rocket mass and acceleration; the right-hand side is therefore the force acting on the rocket, generally referred to as *thrust*

$$\text{Thrust} = Ve \left(-\frac{dM}{dt} \right) \qquad 7\text{-}4$$

Thrust is positive, since the differential quotient is positive; that is, rocket mass M decreases with time. With Ve in units of meters per second and the time rate of mass decrease in units of kilograms per seconds, thrust will be expressed in newtons (N), since by definition

$$N = m \ kg/s^2$$

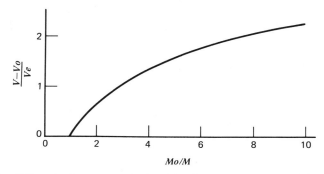

Figure 7.2 Eq. 7-2, velocity-mass relationship for a rocket not subject to external forces.

Note that the right-hand side of Eq. 7-3 is the product of a mass $-dM$ and the acceleration of that mass from zero in the combustion chamber to Ve during a time interval dt. What force does this product represent?

7.2 VERTICAL POWERED FLIGHT
Figure 7.3 helps to fix terminology and notation:

Mo = total rocket mass at lift-off
Mp = propellant mass at lift-off
V = rocket velocity, positive upward
z = rocket elevation above launching pad, positive upward
Mbo = rocket mass at burnout
zbo = rocket elevation at burnout
Vbo = rocket velocity at burnout
tbo = time at burnout, measured from ignition
Vc = rocket velocity during coasting period

At lift-off, time t and velocity V are zero. Altitude above the lift-off point is denoted by z. The total mass of the rocket at lift-off is Mo, and

$$Mo = Mp + Ms + Me + Mpl$$

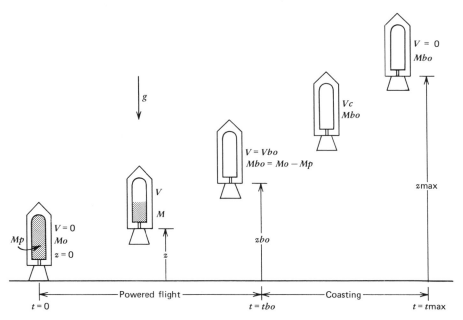

Figure 7.3 Rocket in vertical flight.

and consists of propellant, Mp, structural elements, Ms, engine and accessories, Me, and finally, the payload, Mpl. Powered flight lasts for time tbo. At the instant the last piece of propellant burns the altitude, velocity and mass of the rocket are zbo, Vbo, and Mbo, respectively. After burnout the rocket coasts under the influence of the decelerating force of gravity and ultimately comes to rest at time $tmax$, where the altitude is $zmax$. The latter two quantities, the maximum altitude and the time required to reach it, are of major interest.

During powered flight the rocket is subjected to three forces,

$$M \frac{dV}{dT} = \text{thrust} - \text{drag} - \text{weight} \qquad \text{7-5}$$

Drag and weight act oppositely to thrust. Thrust is given by Eq. 7-4.

$$\text{Weight} = Mg, \qquad \text{where } g = go\left(\frac{R}{R+z}\right)^2 \qquad \text{7-6}$$

in which g is the local gravitational acceleration, go is the surface value, and R is the planet radius.

$$\text{Drag} = \frac{1}{2}\rho V^2 C_d A$$

in which ρ is the local atmospheric density, C_d is a coefficient that depends on rocket shape and Reynold's number, and A is the frontal area.

The full equation, Eq. 7-5, cannot be integrated in closed form. Numerical integration is required, taking small increments in time and proceeding in a stepwise manner to calculate the changes in V and z. Since the drag force depends on rocket size and shape, Eq. 7-5 cannot be integrated for the general case. We wish to deal with rockets in general, and we simplify Eq. 7-5 to

$$M \frac{dV}{dt} = - Ve \frac{dM}{dt} - Mgo \qquad \text{7-7}$$

by neglecting drag and the variation in the local gravitational attraction. (These approximations offset each other.) Two rocket parameters are now introduced;

$$\pi = \text{propellant loading} = \frac{Mp}{Mo} \qquad \text{7-8}$$

$$\beta = \text{burning rate} = -\frac{1}{Mp} \frac{dM}{dt} \qquad \text{7-9}$$

Propellant loading is simply the fraction of total lift-off mass devoted to propellant. Burning rate is the fraction of initial propellant mass Mp that burns in unit time. For example, $\beta = .056$ (1/s) means that 5.6% of Mp will be burned every second. The rocket mass M as a function of time t is now

$$M = Mo - \beta Mpt \qquad \text{7-10}$$

Equation 7-7 can now be integrated to give

$$V = Ve \ln\left(\frac{Mo}{M}\right) - got$$

$$z = \frac{Ve}{\pi\beta} \; [(1 - \pi\beta t)\ln(1 - \pi\beta t) + \pi\beta t] - \tfrac{1}{2}go(t^2) \qquad \text{7-11}$$

At burnout $t = 1/\beta$, and $M = Mo - Mp$. Making these substitutions in Eqs. 7-11,

$$Vbo = Ve \ln\left(\frac{1}{1-\pi}\right) - \frac{go}{\beta} \qquad \text{7-12}$$

$$zbo = \frac{Ve}{\beta}\left[\left(\frac{1}{\pi} - 1\right) \ln(1 - \pi) + 1\right] - \tfrac{1}{2}\left(\frac{go}{\beta^2}\right) \qquad \text{7-13}$$

are the velocity and altitude at burnout.

7.3 VERTICAL FLIGHT: COASTING
Following burnout the rocket coasts, and

$$\frac{dV}{dt} = -g$$

If g is set equal to the surface value go, then by integration

$$zmax = \frac{Ve^2}{2go} \left[\ln\frac{1}{1-\pi}\right]^2 + \frac{Ve}{\beta}\left[1 + \frac{1}{\pi}\ln(1 - \pi)\right] \qquad \text{7-14}$$

and

$$tmax = \frac{Ve}{go}\ln\left(\frac{1}{1-\pi}\right) \qquad \text{7-15}$$

The time required to reach maximum altitude is independent of the burning rate β when the local variation in the gravitational acceleration is ignored.

If the local variation in g is taken into account, the equation of motion during the coasting time is

$$\frac{dV}{dt} = V\frac{dV}{dz} = -go\left(\frac{R}{R+z}\right)^2$$

which will integrate to

$$V^2 = Vbo^2 + 2\,go(R^2)\left[\frac{1}{R+z} - \frac{1}{R+zbo}\right]. \qquad \text{7-16}$$

Setting $V = 0$ gives the maximum altitude,

$$zmax = \frac{1}{\dfrac{1}{R+zbo} - \dfrac{Vbo^2}{2\,go(R^2)}} - R \qquad \text{7-17}$$

and the time to maximum altitude can be found by setting $V = dz/dt$ in Eq. 7-16 and integrating between zbo and $zmax$. The result will be the coasting time, which should be added to tbo in order to arrive at $tmax$.

The equations for Vbo, zbo, $zmax$ and $tmax$ involve π in an awkward manner. To facilitate hand calculations, the several expressions that include π are tabulated at the end of this chapter (Table 7.5). It is recommended that the reader examine these indices of rocket performance in order to acquire some understanding of the separate influences of the two rocket descriptors, β and π. It is evident from Eq. 7-14 that for the case of $g = go$ during the coasting period, $zmax$ increases with Ve. It is to be expected that Eq. 7-17 would show the same trend. To reach high altitude, its desirable to use a propellant combination that gives large Ve.

The propellant loading parameter π can take on all values between zero and unity. However, equating thrust to rocket weight at lift-off, it is at once evident that the burning rate β has a practical lower limit,

$$\beta_{min} = \frac{go}{Ve\,\pi} \qquad \text{7-18}$$

If β is less than β_{min}, the rocket will remain on the launching pad until burning reduces the rocket weight to something less than the thrust.

7.4 THE ESCAPE VELOCITY

When the denominator in Eq. 7-17, the expression for $zmax$ taking into account variation of g with distance from the planet, is zero, $zmax$ will become infinite. This means the rocket will escape from the earth's gravitational field, ignoring the influences of other attracting bodies. The burnout velocity required for escape is

$$Vbo, \text{ escape} = \sqrt{\frac{2\,go(R^2)}{R + zbo}} \qquad \text{7-19}$$

To escape from the surface of the earth ($zbo = 0$), taking $R = 6372$ km, the rocket, or any object projected vertically, must have a velocity equal to 11 190 m/s, ignoring the drag forces, which are only effective during the first few miles of travel but are not insignificant. On the other hand, if burnout occurs at an altitude of one earth radius, the escape velocity is reduced to 7 894 m/s.

7.5 RESTRICTIONS ON β AND π IMPOSED BY ACCELERATIONS

The performance of a rocket in vertical flight is influenced by four parameters:

go = surface gravitational acceleration
Ve = exhaust velocity
β = burning rate
π = propellant loading

(To these could be added the planet radius R.) The location in the universe fixes the value of go. The chemical composition of the propellants and their proportions in the reactant mixture of the engine fix Ve, which is to be calculated by the methods of thermodynamics and will be discussed momentarily. This leaves β and π at the disposal of the rocket designer.

It should not be assumed that β and π can take on any values we choose. Practical considerations and safety requirements impose limitations on the minimum acceleration at lift-off for earth launchings and the maximum acceleration at burnout. At lift-off the rocket mass is large, the net force acting (thrust minus weight) is small, the initial velocity is zero, and the velocity increases slowly with time and altitude. For earth launchings, the rocket is at the mercy of crosswinds. When tilted out of the vertical, the thrust and weight forces form a couple that increases the tilt. These considerations place a lower limit of around $0.25G$ on lift-off acceleration; that is, engine thrust must be at least 1.25 times lift-off weight.

At burnout the rocket mass is small, and the net force acting is large, as is the acceleration. To maintain structural integrity, protect instrumentation and safeguard human life, the upper limit on burnout acceleration is around 10G. These acceleration limits transfer into statements about β and π as follows:

Denoting acceleration by a, from Eq. 7-7, we have

$$a = \beta Ve \frac{Mp}{M} - go$$

The lift-off acceleration, with $M = Mo$, is

$$ao = \beta\pi Ve - go$$

and if ao must be not less than $0.25go$

$$(\beta\pi)_{min} Ve = 1.25go \qquad\qquad 7\text{-}20$$

At burnout, $M = Mo - Mp$, and the acceleration is

$$abo = \beta\frac{\pi}{1-\pi} V_e - go$$

With abo less than $10go$,

$$\left(\beta\frac{\pi}{1-\pi}\right)_{max} Ve = 11go \qquad\qquad 7\text{-}21$$

For a selected value of Ve, Eqs. 7-20 and 7-21 may be plotted on a β versus π plane, as in Fig. 7.4. Lines of constant $z$max have been traced out inside the allowable operating

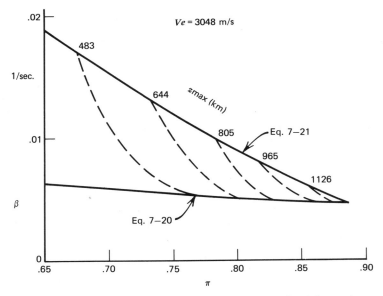

Figure 7.4 Allowable operating range for a vertical rocket, fired from the earth, with lift-off acceleration greater than .25 G, and burnout acceleration less than 10 G.

region. The best altitude performance occurs at the intersection of Eqs. 7-20 and 7-21, for which the coordinates are

$$\pi = 1 - \frac{ao + go}{abo + go}$$

$$\beta = \frac{1}{Ve} \frac{(ao + go)(abo + go)}{abo - ao} \qquad \text{7-22}$$

For a rocket fired vertically from earth, the largest attainable values for $z$max, calculated from Eq. 7-14, which assumes local acceleration equal to surface gravitational acceleration during the long coasting period, and the corresponding $t$max, time from lift-off to maximum altitude, vary with Ve as follows:

Ve (m/s)	$z$max (km)	$t$max (s)
2440	800	542
3050	1250	677
3660	1800	812

7.6 THE NOZZLE EXIT VELOCITY

A rocket engine is sketched in Fig. 7.5. Fuel and oxidizer are pumped into the combustion chamber at steady rates of flow, burn at pressure Pc and temperature Tc, pass through a converging-diverging nozzle, and flow out through the exit section at pressure Pe, temperature Te, and with velocity Ve, relative to the nozzle.

For a control volume that intercepts the incoming reactants and the combustion chamber, the steady flow energy equation for adiabatic combustion at constant pressure is

$$H_r = H_p(Tc)$$

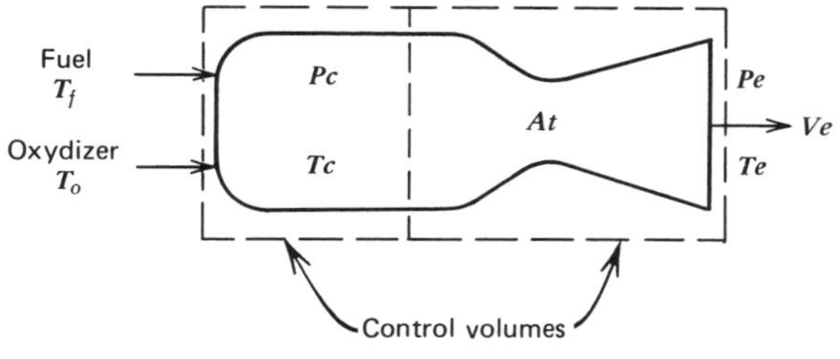

Figure 7.5 Rocket motor, burning liquid propellents.

where H_r and H_p denote the enthalpies of the reactants and products, respectively. For given fuel/oxygen ratio the composition of the product gas mixture can be set down at once and the energy equation solved for Tc by the methods outlined and discussed in Chapter 2.

Taking a control volume in Fig 7.5 which extends from the combustion chamber to the nozzle exit area, the steady flow energy equation for adiabatic flow can be written as

$$h(Tc) = h(Te) + \tfrac{1}{2}Ve^2$$

which ignores the velocity in the combustion chamber by assuming it to be negligibly small. The exit velocity Ve is then

$$Ve = \sqrt{2[h(Tc) - h(Te)]} \qquad\qquad 7\text{-}23$$

Since $Ve^2/2$ is the energy of the product gas mixture on a *mass* basis, the enthalpy terms must be stated in similar units. Alternatively, we can write for Ve,

$$Ve = \sqrt{2\frac{\Sigma\, n_i[h_i(Tc) - h_i(Te)]}{\Sigma\, n_i mw_i}} \qquad\qquad 7\text{-}24$$

In Eq. 7-24, n_i and mw_i denote mole numbers and corresponding molecular weights for the products gases, and the enthalpy terms are stated in terms of moles.

With Tc known, the problem is to compute Te, which can be accomplished in three ways, as outlined in Chapter 3. The simplest is

$$Te = Tc\left(\frac{Pe}{Pc}\right)^{(k-1)/k} \qquad\qquad 7\text{-}25$$

which assumes that the specific heat ratio k is constant. A refinement of Eq. 7-25 is

$$Te = Tc\left(\frac{Pe}{Pc}\right)^{(k_m-1)/k_m} \qquad\qquad 7\text{-}26$$

where k_m is an average value for k based on the values at Tc and Te from Eq. 7-25. The exact calculation of Te is based on the condition of zero entropy change between combustion chamber and nozzle exit section, that is,

$$S(Pc, Tc) = S(Pe, Te) \qquad\qquad 7\text{-}27$$

which can be solved for Te by the method discussed in Chapter 3.

TABLE 7.1
$C_{10}H_{22}$(LIQUID) + Y O_2(LIQUID)
Y = moles O_2/mole fuel; Pc = 30 atm; Pe = 1 atm; Ve(m/s)

		Eq. 7-25		Eq. 7-26		Eq. 7-27	
Y	Tc	Te	Ve	Te	Ve	Te	Ve
12.40	4682	2732	2797	2699	2821	2703	2818
13.95	5130	3118	2832	3089	2852	3091	2851
15.50	5518	3472	2846	3446	2864	3448	2862
17.05	5269	3280	2767	3252	2786	3254	2785
18.60	5043	3108	2695	3078	2715	3080	2713

TABLE 7.2
H_2(LIQUID + Y O_2(LIQUID)
Y = moles O_2/mole H_2; Pc = 20 atm; Pe = 1 atm; Ve(m/s)

		Eq. 7-25		Eq. 7-27	
Y	Tc	Te	Ve	Te	Ve
0.7	4271	2654	2886	2684	2861
0.6	4541	2858	3066	2886	3041
0.5	4851	3096	3289	3122	3265
0.4	4215	2586	3358	2626	3318
0.3	3484	2030	3405	2080	3349
0.25	3071	1716	3418	1783	3341

What types of results do these three methods produce? The question has no general answer, of course. Table 7.1 gives some calculation results for Te and Ve by the three methods for kerosene/liquid oxygen, and Table 7.2 shows results for liquid hydrogen/liquid oxygen for two of the methods. Clearly, all three methods give answers that are very close to one another. In using Eq. 7-26, k_m was derived as a simple arithmetic mean. The combustion chamber temperature Tc and velocity Ve both reach their peak values at chemically correct mixture ratio (Y = 15.5). On the other hand, in the calculations in Table 7.2, Tc peaks at chemically correct mixture (Y = 0.5), but Ve peaks at Y = 0.3, using Eq. 7-27. Why do these two combustion systems behave differently with regard to Ve?

7.7 THE REQUIRED MASS FLOW RATE

The normal range for combustion chamber pressure is 20 to 30 atm. Nozzle exit section pressure is 1 atm or less. With pressure ratios of this magnitude, gases flowing through the nozzle reach the acoustic velocity at the throat section. The mass flow rate can be computed by borrowing a result from fluid mechanics.

$$\dot{M} = \Gamma A t P c \sqrt{\frac{MW}{RTc}}$$ 7-28

where $\dot{M}$ is the mass flow rate, At is the nozzle throat area, Pc and Tc are the combustion chamber pressure and temperature, R is the universal gas constant, MW is the molecular weight, and Γ is a function of the specific heat ratio k.

$$\Gamma = \sqrt{k}\left(\frac{2}{k+1}\right)^{\frac{k+1}{2(k-1)}}$$

which is not a strong function of k. Since

$$\dot{M} = \beta Mp$$ 7-29

Equations 7-28 and 7-29 provide a relationship between Mp, the propellant mass; β, the burning rate; and $\dot{M}$, the mass flow rate.

Alternatively, the mass flow rate through the nozzle can be determined by discovering the pressure and temperature conditions at the throat of the nozzle. For the mass flow rate,

$$\dot{M} = \rho t A t V t$$ 7-30a

where ρt and Vt are the density and gas velocity at the throat. The throat velocity must be the acoustic velocity, which, for an ideal gas or mixture of ideal gases, is

$$Vt = V_{\text{acoustic}} = \sqrt{\frac{kRTt}{MW}}$$ 7-30b

where Tt denotes absolute temperature at the throat. The stream velocity Vs is

$$Vs = \sqrt{2[h(Tc) - h(Tt)]}$$

and the condition to be satisfied is

$$Vs = Vt$$ 7-30c

This can be accomplished with a half-interval search routine. Start with the throat pressure Pt in the middle of its allowable range,

$$Pt = \frac{Pc + Pe}{2}$$

and calculate the corresponding throat temperature Tt from the exact relationship for isentropic flow

$$S(Pc, Tc) = S(Pt, Tt)$$

after which Vs and Vt are calculated and compared. As a gas moves through a nozzle, the stream velocity increases (if there are no shock fronts), and the local acoustic velocity decreases. This provides the logic with which to direct the search for the throat pressure and temperature.

7.8 THRUST AND SPECIFIC IMPULSE

With mass flow rate $\dot{M}$ and exhaust velocity Ve known, the *thrust* produced by the motor is

$$\text{Thrust} = \dot{M}Ve \qquad\qquad 7\text{-}31$$

and has units of force.

The *specific impulse*, Isp, is the amount of thrust force developed per unit mass of propellant burned per unit time, that is,

$$Isp = \frac{thrust}{\dot{M}} = Ve \ (\text{N/kg/s}) \qquad\qquad 7\text{-}32$$

In the English unit system

$$Isp = \frac{thrust}{\dot{M}} = \frac{Ve}{gc} (\text{lbf/lbm/sec})$$

(Specific impulse is sometimes reported in the English system in units of seconds. This is improper and results from substituting the local gravitational acceleration g for the dimensional constant gc.)

7.9 SOME PERFORMANCE FIGURES FOR VERTICAL FLIGHT

Tables 7.3 and 7.4 list some of the calculated results for the following rocket:

$$Mo \doteq 13\ 605 \text{ kg}$$
$$Mp = 12\ 245 \text{ kg}$$
$$At = 0.056 \text{ m}^2$$
$$Pc = 20 \text{ atm}$$
$$Pe = 1 \text{ atm}$$

TABLE 7.3
$C_{10}H_{22}$/LIQUID O_2 ROCKET

Moles O_2 / Moles Fuel	tbo (s)	Vbo (m/s)	zbo (km)	zmax (km)	Thrust (kN)
20.15	191	3928	179	1131	162
18.60	195	4029	188	1199	162
17.05	200	4138	198	1278	162
15.50	205	4257	209	1368	163
13.95	206	4228	208	1347	161
12.40	205	4169	203	1303	160
10.85	203	4066	195	1231	159

TABLE 7.4
LIQUID H_2/LIQUID O_2 ROCKET

Moles O_2 / Moles H_2	tbo (s)	Vbo (m/s)	zbo (km)	zmax (km)	Thrust (kN)
0.2	297	4905	318	2006	140
0.3	291	4993	322	2090	144
0.4	282	4957	314	2045	146
0.5	274	4887	303	1967	147
0.6	257	4551	263	1642	146
0.7	242	4274	233	1409	146

for two systems, one fueled with $C_{10}H_{22}$/liquid O_2 and the other with liquid H_2/liquid O_2. In both tables, g is constant during the burning period but varies during the coasting period. In Fig. 7.6 the altitude performances are plotted against equivalence ratios. The ordinate is extended to zero to emphasize the substantial distance rockets travel during coasting.

Note that the kerosene/oxygen fuel system produces maximum altitude at chemically correct mixture, while the hydrogen/oxygen system shows continuing gains in maximum altitude as the mixture becomes progressively richer in hydrogen.

These calculations have been carried through without regard for a question of enormous signficance in rocket predictions. Will a particular fuel/oxygen mixture burn at a specified combustion chamber pressure at the required rate of propellant mass flow rate? If the answer is no, then the predictions are meaningless. Unfortunately, thermodynamics is of no help whatsoever in the search for an answer to the question. Nor, for that matter, has chemical kinetics reached the stage at which it can supply definitive answers. The problem of determining burning rate from first principles is too complicated to be solved on the basis of present partial understanding of reaction rates. Rocket predictions must be confirmed by experiment. However, our ability to extrapolate small rocket performance to large systems seems to be successful.

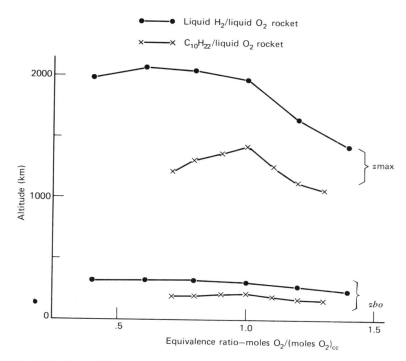

Figure 7.6 Altitude performance of two rockets, fired vertically from the earth. Mo = 13 605 kg Mp = 12 245 kg $(\pi = .9)$, At = .056 m^2, Pc = 20, Pe = 1 atm. zbo calculated with constant g; $z$max calculated with variable g.

7.10 RETRO—ROCKETS

A retro-rocket decelerates a space vehicle. The notation is illustrated in Fig. 7.7. At some point during the coasting period the position above the surface is zs, and the velocity is Vs; z is measured positive upward, and velocity is positive downward. The velocity Vig and the altitude zig at the instant the retro-rocket engine is ignited are related by

$$Vig^2 = Vs^2 + 2go(zs - zig) \qquad 7\text{-}33$$

if the local variation in gravitational acceleration is ignored, or by

$$Vig^2 = Vs^2 + 2go(R^2)\left[\frac{1}{R + zig} - \frac{1}{R + zs}\right] \qquad 7\text{-}34$$

when the local variation is included. R is the planet radius.

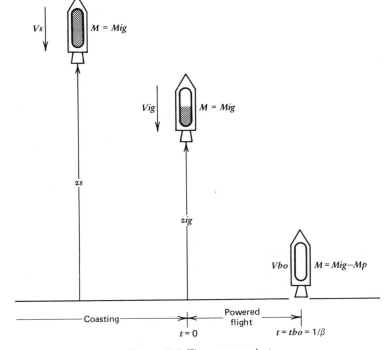

Figure 7.7 The retrorocket.

During powered flight, the motion is governed by the equation

$$M\frac{dV}{dt} = Mgo + Ve\frac{dM}{dt}$$

7-35

which ignores drag and assumes $g = go$ as an approximation. Defining π and β as before,

$$\pi = \frac{Mp}{Mig}, \qquad \beta = -\frac{1}{Mp}\frac{dM}{dt}$$

Equation 7-35 integrates to give the following

$$Vig = Vbo - Ve\ln(1 - \pi) - \frac{go}{\beta}$$

7-36

$$zig = \frac{Vig}{\beta} - \frac{1}{2}\frac{go}{\beta^2} - \frac{Ve}{\beta}\left[\left(\frac{1}{\pi} - 1\right)\ln(1 - \pi) + 1\right]$$

7-37

where Vbo is the vehicle velocity at touchdown, when all the propellant has been burned. Once zs and Vs have been assigned, Eq. 7-33 or Eq. 7-34 defines the relation between zig and Vig, so that Eqs. 7-36 and 7-37 give the relationship between β and π for given (zig, Vbo) or (Vig, Vbo).

At ignition, the vehicle and contents are subjected to a sudden jolt, and some upper limit must be imposed on the acceleration at ignition, aig, given by

$$aig = \beta\pi Ve - go \qquad\qquad 7\text{-}38$$

taken positive. When acceleration at ignition becomes a controlling parameter in the deceleration process, then the possible variations in β and π are defined by the pair (aig, Vbo).

Figure 7.8 illustrates a moon landing for a rocket with Ve = 3048 m/s, and touchdown velocity Vbo = 3.05 m/s. On the moon, go = 1.62 m/s^2. Equation 7-34 is shown and represents the velocity of approach. If there is no retro-rocket, the vehicle will crash-land at about 2340 m/s, which is the escape velocity from the moon. Ignition must occur at some point on the line defined by Eq. 7-34. As shown,

$$\begin{aligned}
\text{If } aig &= 1 \text{ G}, & zig &= 237 \text{ km} \\
aig &= 2 \text{ G}, & zig &= 116 \text{ km} \\
aig &= 3 \text{ G}, & zig &= 76 \text{ km}
\end{aligned}$$

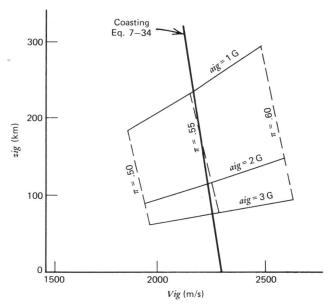

Figure 7.8 A soft moon landing for a rocket with Ve = 3048 m/s, Vbo = 3.05 m/s.

As ignition is delayed, the ignition acceleration increases. Note in Fig. 7.8 that the line for $\pi = 0.55$ virtually coincides with Eq. 7-34. As a consequence, the burning rate β is fixed by *aig*, from Eq. 7-38

$$\beta = \frac{aig + go}{\pi Ve}$$

so that for *aig* = 1 G, for example,

$$\beta = \frac{9.8 + 1.62}{0.55 \times 3048} = 0.0068 \ (1/s)$$

or a burning time of $1/0.0068 = 147$ s. If *aig* is increased, the burning rate also increases, and the burning time decreases.

Some appreciation for the split-second timing required for soft moon landings can be gained from a simple calculation. Near the moon, the velocity and altitude under coasting vary according to

$$\frac{d(Vs)}{d(zs)} = -\frac{go}{Vs} = \frac{1.62 \ m/s^2}{2286 \ m/s}$$

$$= 0.71 \ m/s/km$$

According to Eq. 7-36, for a given rocket, that is, for fixed values of the three rocket parameters π, β, and Ve,

$$d(Vig) = d(Vbo)$$

This means that if ignition is delayed or advanced by 1 km, the touchdown velocity, *Vbo*, will change by 0.71 m/s, which is large compared to 3 m/s, the desirable order of magnitude for *Vbo*. Note further that when moving at a velocity of 2286 m/s, the vehicle travels 1 km in 0.44 seconds! To achieve a predetermined touchdown velocity, the instant for ignition must be correct within less than one-half second.

Equation 7-33 relates altitude above the moon and velocity for an object in free fall under the sole influence of the attractive force of the moon. If the rocket has been fired from the earth and is approaching the moon for a soft landing, then Eq. 7-39 (see below) is more appropriate as the relation between altitude and velocity.

7.11 EARTH TO MOON MISSIONS

The techniques and equations developed in the previous sections of this chapter can be applied to an extended rocket flight from earth to moon. Inclusion of the return trip is a more ambitious problem, which can be treated with the methods to be outlined. The orbital mechanics will be substantially simplified to make the analysis more tractable. The performance of two-stage rockets is examined briefly.

Consider the following question: For a given rocket motor and propellant combination, what is the *minimum* propellant mass required to place a payload of given mass on the moon? As stated, the question has no answer; it is far too general. For one thing, the maximum transit time for the flight must be specified, so we first examine calculation of the flight time.

The real motion of an object projected from the earth in such a way as to land on the moon is quite complicated. We shall simplify the system of earth and moon by assuming that both remain stationary during the trip, and by ignoring the earth's rotation. (The earth's rotation may be utilized, of course, and the advantage that can be gained is discussed at the end of this chapter.) Referring to Fig. 7.9, we assume the earth and moon to be separated by 386 160 km, and we shall use the following astronomical dimensions:

	Surface g	Radius
Earth	$goe = 9.80$ m/s^2	$Re = 6436$ km
Moon	$gom = 1.62$	$Rm = 1737$

With the earth and moon both stationary, there is a "zero-gravity" point where the gravitational attractions are equal and oppositely directed, which is calculated from

$$goe\left(\frac{Re}{x}\right)^2 = gom\left(\frac{Rm}{D-x}\right)^2$$

where x is measured from the earth's center. The distance to the zero-gravity point turns out to be about 54.1 earth radii.

Ignoring drag forces encountered during passage through the earth's atmosphere, the equation for straight-line motion from earth to moon is

$$V\frac{dV}{dx} = -goe\left(\frac{Re}{x}\right)^2 + gom\left(\frac{Rm}{D-x}\right)^2$$

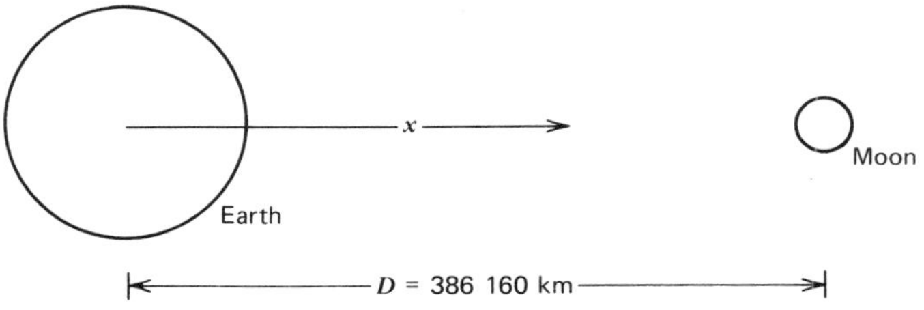

Figure 7.9 A simplified arrangement of stationary moon and earth.

Here V is the velocity toward the moon. The equation integrates to

$$V(\text{m/s}) = \left[126 \times 10^6 \left\{\frac{0.012}{60-r} + \frac{1}{r} - 0.021\right\} + Vzg^2\right]^{1/2} \qquad 7\text{-}39$$

r being the distance from earth's center, expressed in earth radii, and Vzg is the velocity at the zero-gravity point, in units of meters per second.

   An object fired vertically from the earth's surface and aimed at the moon with a velocity 11 100 m/s will just reach the zero-gravity point and remain there indefinitely. Given a nudge toward the moon, it will crash-land at 2260 m/s. The variation of V with log(r) can be found by setting $Vzg = 0$ in Eq. 7-39; this is plotted in Fig. 7.10.

   Now suppose the flight time is specified. Then

$$\text{Time of flight} = Re \int_1^{59.73} \frac{dr}{V} \qquad 7\text{-}40$$

Equation 7-40 ignored small corrections that arise from accelerated flight near the earth and decelerated flight near the moon. These corrections amount to minutes at most; it is assumed that the flight time is measured in days. Equation 7-40 provides the value for Vzg, which is to be found by evaluating the integral for various values of Vzg

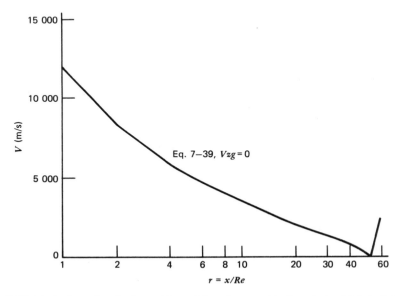

Figure 7.10 Variation of velocity with position for a vertical rocket which just reaches the zero-gravity point.

and employing a systematic routine for changing Vzg so that the integral converges on the required value. (The use of Vzg as a parameter in Eq. 7-39 is a convenience. Also, it is interesting to discover the required value for Vzg for some given flight time.)

A single-stage rocket cannot reach the moon with the propellants now available. The maximum value for Ve is about 3660 m/s. With a burning time of 120 s, the propellant loading required to reach 11 100 m/s burnout velocity can be calculated from Eq. 7-12, which is repeated here

$$Vbo = Ve \ln\left(\frac{1}{1 - \pi}\right) - \frac{goe}{\beta}$$

and turns out to be $\pi = 0.965$. This leaves 3.5% of the lift of mass of the rocket for engine, fuel tanks, structure, and payload. Even if a rocket could be constructed with that mass distribution, the burnout acceleration is prohibitive,

$$abo = \frac{\pi}{1 - \pi} \beta Ve - goe = 85 \text{ G}$$

An earth-to-moon mission requires at least a two-stage rocket. We discuss rockets of that type next.

7.12 TWO-STAGE ROCKETS

The powered flight path of a two-stage rocket is drawn in Fig. 7.11. The maximum and minimum flight times, specified as part of the problem, can be translated into coasting curves, as shown. To reach the moon within the required time bracket, burnout of the second stage must occur inside the "window." Note the costly velocity loss that can occur between first-stage burnout and second-stage ignition. (During powered flight, velocity and altitude show an approximately linear relation. The slope, however, may not be the same for both stages.) For the case drawn in Fig. 7.11, second-stage burnout leaves the rocket outside the window; hence the vehicle will arrive at the moon late, if it gets there at all.

Figure 7.11 describes one of the difficult aspects of the problem, namely, finding a combination of stages that will land the rocket inside the window, and then finding what distribution of propellant mass between the two stages requires the minimum total amount.

It is instructive to examine the increase in burnout velocity that can be achieved with staging. Suppose, for example, that a single-stage rocket has the following distribution of mass at lift-off:

$$Mpl = \text{payload} = 900 \text{ kg}$$
$$Ms = \text{structure and engine} = 2700$$
$$Mp = \text{propellant} = 14\ 400$$
$$Mo = \text{total} = 18\ 000 \text{ kg}$$

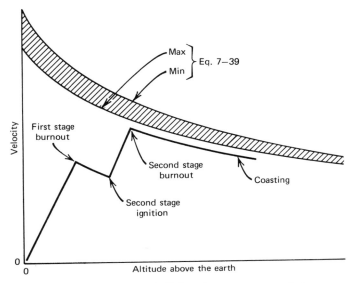

Figure 7.11 Powered flight of a two stage rocket.

Suppose further that $Ve = 3048$ m/s and that the lift-off acceleration is set at $0.25G$. Since $\pi = 14\,400/18\,000 = 0.8$

$$\beta = \frac{ao + go}{\pi Ve} = 0.00503 \; 1/s$$

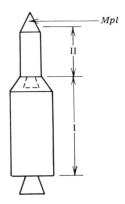

Figure 7.12 A two-stage rocket. Stages are identified by Roman numerals and are numbered in the firing order.

the burnout acceleration will be

$$abo = \beta \, \frac{1}{1-\pi} \, Ve - goe = 5.25 \text{ G}$$

and the burnout velocity is

$$Vbo = Ve \ln \frac{1}{1-\pi} - \frac{goe}{\beta} = 2957 \text{ m/s}$$

Now let the same masses be divided between two stages (see Fig. 7.12):

	I Stage	II Stage	
Mpl	?	900 kg	
Ms	?	?	(Total = 2700)
Mp	?	?	(Total = 14 400)
Mo	18 000	?	

The rocket motor in both stages produces a $Ve = 3048$ m/s; lift-off acceleration is again 0.25 G, and second-stage burnout acceleration is to be 5.25 G. (With these requirements, comparison between single- and two-stage rocket performance is more meaningful.) There are four blanks above, namely, Ms_I, Ms_{II}, Mp_I, and Mp_{II}. We need four equations. Two are

$$Ms_I + Ms_{II} = 2700$$
$$Mp_I + Mp_{II} = 14\ 400$$

The other two could be

$$\frac{Mp_I}{Ms_I} = \frac{Mp_{II}}{Ms_{II}} = \frac{14\ 400}{2700}$$

that is, the mass ratio of propellant/structure is identical for both stages. (A more sophisticated treatment of mass distribution is discussed in a moment.) The performance of the rocket can now be calculated. The variation of second-stage burnout velocity with second-stage propellant mass is drawn in Fig. 7.13, where the curve is observed to be flat, suggesting that mass distribution is not critical, at least for this particular rocket. The mass distribution that achieves maximum second-stage burnout velocity is as follows:

	I Stage	II Stage
Mpl	2984	900
Ms	2371	1067
Mp	12 645	1755
Mo	18 000	2984

Note that the payload of the first stage equals the mass of the second stage at second-stage ignition. The second-stage burnout velocity is calculated from the sum of the velocity increments of each stage

$$Vbo_{II} = \Delta V_I + \Delta V_{II} - g \, \Delta t$$

where Δt is the time interval between first-stage burnout and second-stage ignition, which was ignored in calculating Fig. 7.13. For the two stages:

	I Stage	II Stage
π	0.703	0.588
β	0.00573	0.0141 1/s
ΔV	1987	2006 m/s
	$Vbo_{II} = 3993$ m/s	

The velocity increments of the two stages are almost identical. It can be shown that they are exactly equal when the propellant/structure mass ratio is the same for both stages.

A more sophisticated treatment of mass distribution in a multistage rocket takes note of the fact that propellant mass is proportional to propellant volume and therefore to the cube of some representative length, while structural mass, on the

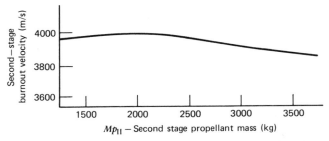

Figure 7.13 Variation of second-stage burnout velocity.

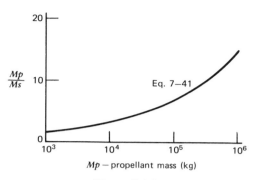

Figure 7.14

other hand, is approximately proportional to the cross-sectional area of members and hence to the square of the same representative length. Consequently,

$$\frac{Mp}{Ms} \sim \frac{L^3}{L^2} \sim L \sim Mp^{1/3}$$

For design purposes, the following is reasonable:

$$\frac{Mp}{Ms} = \frac{Mp^{1/3}}{6.5} \qquad (Mp \text{ in kg}) \qquad 7\text{-}41$$

Figure 7.14, a plot of Eq. 7-41, reveals the advantage of large rockets, since the structural mass required per unit of propellant decreases as propellant mass increases. However, the factor 6.5 in Eq. 7-41 is not firm; the proper value depends on rocket design, shape, and size.

When Eq. 7-41 is employed to distribute mass in a multistage rocket, the whole procedure becomes far more complicated than the example given above. A trial-and-error approach to the search for optimum distribution seems to be the only available strategy.

This brief examination of a two-stage rocket illustrates the increase in burnout velocity achieved with staging — in this case, an increase from 2957 to 3993 m/s.

7.13, GRAVITY TURNS AND CIRCULAR ORBITS

In practice, earth-to-moon missions are accomplished by placing the vehicle in a circular orbit about the earth. At the instant the vehicle passes directly behind the earth, the velocity is boosted. The vehicle departs from the circular orbit and travels to the moon along a semielliptical trajectory. This type of motion is referred to as a Hohmann-type transfer. It has the advantage of the earth's tangential velocity, which

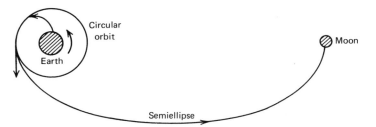

Figure 7.15 Hohmann-type transfer from earth to moon.

amounts to 470 m/s at the equator. Figure 7.15 is a sketch of the circular orbit and the semielliptical transfer to the moon.

Although Hohmann-type transfers generally require the least energy, the calculations are vastly more complicated than those for the simple straight-line trajectory discussed earlier. The velocity of the circular orbit, Vco, and the altitude of the orbit above the earth's surface, zco, are related by

$$Vco = Re\left(\frac{go}{Re + zco}\right)^{1/2} \qquad\qquad 7\text{-}42$$

which is obtained by equating the earth's gravitational attraction and the force arising from centripetal acceleration. However, in order to place a vehicle in circular orbit, the rocket must perform a gravity turn, as illustrated in Fig. 7.16. Instead of Eq. 7-7, the motion during powered flight is governed by

$$M\frac{dV}{dt} = -Ve\frac{dM}{dt} - Mgo\cos\theta \qquad\qquad 7\text{-}43$$

where θ is the angle between flight direction and the vertical.

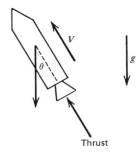

Figure 7.16 Powered flight during a gravity turn.

The motion is now two-dimensional, and Eq. 7-43 must be integrated numerically in both dimensions, since the variation of angle θ with position and/or time is not known. The major problem is deciding where to begin the gravity turn during the powered flight in order to coast into a circular orbit with Vco and zco values that satisfy Eq. 7-42. After burnout, the first term on the right-hand side of Eq. 7-43 drops out, but the $\cos \theta$ factor persists during the coasting period.

7.14 SUMMARY

1. During powered upward flight, the gain in altitude and velocity of the rocket depends on:
   a. go, the surface gravitational acceleration of the local planet.
   b. Ve, the exhaust velocity (measured relative to the rocket) at the nozzle exit section.
   c. π, the ratio of propellant mass to total stage mass at ignition.
   d. β, the mass burning rate.
2. Velocity Ve depends on the reactants, the reactant mixture, combustion chamber pressure, and nozzle geometry.
3. Large burning rates reduce the amount of energy expended in lifting unburned propellant.
4. Large burning rates produce large accelerations. For safe operation, accelerations must be limited.
5. The problem of determining the minimum propellant mass required for an earth-to-moon mission can be studied at two levels of complexity:
   a. A crash landing,
   b. A soft landing,
   and at two further sublevels:
   c. Zero ignition delay time between stages,
   d. Finite ignition delay time between stages,
   and at two still further sublevels:
   e. Propellant/structure mass the same for all stages,
   f. Propellant/structure mass according to Eq. 7-41.
   The mission can be accomplished with one-stage and two-stage rockets by allowing Ve to assume large and presently unattainable values.
   The interesting feature of the minimum propellant problem is the fact that the logic that guides the search is obscure.

TABLE 7.5
FUNCTIONS OF π ENCOUNTERED IN ROCKET CALCULATIONS

π	$\dfrac{\pi}{1-\pi}$	$\ln\left(\dfrac{\pi}{1-\pi}\right)$	$\left(\dfrac{1}{\pi}-1\right)\ln(1-\pi)+1$	$\left[\ln\left(\dfrac{1}{1-\pi}\right)\right]^2$	$1+\dfrac{1}{\pi}\ln(1-\pi)$
0.10	0.111	0.105	0.052	0.011	-0.054
0.11	0.124	0.117	0.057	0.014	-0.059
0.12	0.136	0.128	0.063	0.016	-0.065
0.13	0.149	0.139	0.068	0.019	-0.071
0.14	0.163	0.151	0.074	0.023	-0.077
0.15	0.176	0.163	0.079	0.026	-0.083
0.16	0.190	0.174	0.085	0.030	-0.090
0.17	0.205	0.186	0.090	0.035	-0.096
0.18	0.220	0.198	0.096	0.039	-0.103
0.19	0.235	0.211	0.102	0.044	-0.109
0.20	0.250	0.223	0.107	0.050	-0.116
0.21	0.266	0.236	0.113	0.056	-0.122
0.22	0.282	0.248	0.119	0.062	-0.129
0.23	0.299	0.261	0.125	0.068	-0.136
0.24	0.316	0.274	0.131	0.075	-0.143
0.25	0.333	0.288	0.137	0.083	-0.151
0.26	0.351	0.301	0.143	0.091	-0.158
0.27	0.370	0.315	0.149	0.099	-0.166
0.28	0.389	0.329	0.155	0.108	-0.173
0.29	0.408	0.342	0.161	0.117	-0.181
0.30	0.429	0.357	0.168	0.127	-0.189
0.31	0.449	0.371	0.174	0.138	-0.197
0.32	0.471	0.386	0.180	0.149	-0.205
0.33	0.493	0.400	0.187	0.160	-0.214
0.34	0.515	0.416	0.193	0.173	-0.222
0.35	0.538	0.431	0.200	0.186	-0.231
0.36	0.562	0.446	0.207	0.199	-0.240
0.37	0.587	0.462	0.213	0.213	-0.249

−0.258	0.229	0.220	0.478	0.613	0.38
−0.267	0.244	0.227	0.494	0.639	0.39
−0.277	0.261	0.234	0.511	0.667	0.40
−0.287	0.278	0.241	0.528	0.695	0.41
−0.297	0.297	0.248	0.545	0.724	0.42
−0.307	0.316	0.255	0.562	0.754	0.43
−0.318	0.336	0.262	0.580	0.786	0.44
−0.329	0.357	0.269	0.598	0.818	0.45
−0.340	0.380	0.277	0.616	0.852	0.46
−0.351	0.403	0.284	0.635	0.887	0.47
−0.362	0.428	0.292	0.654	0.923	0.48
−0.374	0.453	0.299	0.673	0.961	0.49
−0.386	0.480	0.307	0.693	1.000	0.50
−0.399	0.509	0.315	0.713	1.041	0.51
−0.411	0.539	0.322	0.734	1.083	0.52
−0.425	0.570	0.330	0.755	1.128	0.53
−0.438	0.603	0.339	0.777	1.174	0.54
−0.452	0.638	0.347	0.799	1.222	0.55
−0.466	0.674	0.355	0.821	1.273	0.56
−0.481	0.712	0.363	0.844	1.326	0.57
−0.496	0.753	0.372	0.867	1.381	0.58
−0.511	0.795	0.380	0.892	1.439	0.59
−0.527	0.840	0.389	0.916	1.500	0.60
−0.544	0.887	0.398	0.942	1.564	0.61
−0.561	0.936	0.407	0.968	1.632	0.62
−0.578	0.989	0.416	0.994	1.703	0.63
−0.596	1.044	0.425	1.022	1.778	0.64
−0.615	1.102	0.435	1.050	1.857	0.65
−0.635	1.164	0.444	1.079	1.941	0.66
−0.655	1.229	0.454	1.109	2.030	0.67
−0.676	1.298	0.464	1.139	2.125	0.68
−0.697	1.372	0.474	1.171	2.226	0.69
−0.720	1.450	0.484	1.204	2.333	0.70
−0.743	1.532	0.494	1.238	2.448	0.71

TABLE 7.5
FUNCTIONS OF π ENCOUNTERED IN ROCKET CALCULATIONS

π	$\dfrac{\pi}{1-\pi}$	$\ln\left(\dfrac{1}{1-\pi}\right)$	$\left(\dfrac{1}{\pi}-1\right)\ln(1-\pi)+1$	$\left[\ln\left(\dfrac{1}{1-\pi}\right)\right]^2$	$1+\dfrac{1}{\pi}\ln(1-\pi)$
0.72	2.571	1.273	0.505	1.620	−0.768
0.73	2.704	1.309	0.516	1.714	−0.794
0.74	2.846	1.347	0.527	1.815	−0.820
0.75	3.000	1.386	0.538	1.922	−0.848
0.76	3.167	1.427	0.549	2.037	−0.878
0.77	3.348	1.470	0.561	2.160	−0.909
0.78	3.545	1.514	0.573	2.293	−0.941
0.79	3.762	1.561	0.585	2.436	−0.975
0.80	4.000	1.609	0.598	2.590	−1.012
0.81	4.263	1.661	0.610	2.758	−1.050
0.82	4.556	1.715	0.624	2.941	−1.091
0.83	4.882	1.772	0.637	3.140	−1.135
0.84	5.250	1.833	0.651	3.358	−1.182
0.85	5.667	1.897	0.665	3.599	−1.232
0.86	6.143	1.966	0.680	3.866	−1.286
0.87	6.692	2.040	0.695	4.162	−1.345
0.88	7.333	2.120	0.711	4.495	−1.409
0.89	8.091	2.207	0.727	4.872	−1.480
0.90	9.000	2.303	0.744	5.302	−1.558
0.91	10.111	2.408	0.762	5.798	−1.646
0.92	11.500	2.526	0.780	6.379	−1.745
0.93	13.286	2.659	0.800	7.072	−1.859
0.94	15.666	2.813	0.820	7.915	−1.993
0.95	19.000	2.996	0.842	8.974	−2.153
0.96	23.999	3.219	0.866	10.361	−2.353
0.97	32.332	3.507	0.892	12.296	−2.615
0.98	48.998	3.912	0.920	15.304	−2.992
0.99	98.991	4.605	0.953	21.207	−3.652

REFERENCES

The following texts contain comprehensive discussions of the theory and operation of combustion chambers, nozzles, compressors, and turbines. Both discuss rocket propulsion and rocket missions.

P. G. Hill and C. R. Peterson; *Mechanics and Thermodynamics of Propulsion*, Addison-Wesley Publishing Co., Reading, Massachusetts, 1965.
D. G. Sheperd, *Aerospace Propulsion*, American Elsevier Publishing Co., New York, 1972.

A broad treatment of rocket propellants, including some discussion of nonequilibrium effects, will be found in

I. Glassman and R. F. Sawyer; *The Performance of Rocket Propellants*, AGARDograph No. 129, Circa Publications, Pelham, New York, 1970.

Two handbooks are available that cover rocket propulsion and performance:

A. R. Koelle (Editor) *Handbook of Astronautical Engineering*, McGraw-Hill, New York, 1961.
H. S. Seifert (Editor) *Space Technology*, John Wiley & Sons, New York, 1959.

PROBLEMS

7.1 Suppose a $C_{10}H_{22}$/liquid oxygen rocket has the following mass distribution at lift-off:

$$\text{Pay load} = Mpl = 1000 \text{ kg}$$
$$\text{Structure and engine} = Ms + Me = 2000$$
$$\text{Propellant} = Mp = 17\,000$$
$$\text{Total} = Mo = 20\,000 \text{ kg}$$

Let Pc = 30 atm and Pe = 1 atm, and in addition pick a nozzle throat area that will give a burnout acceleration of 9 G, using a chemically correct propellant ratio.

Now observe how the following performance parameters:

(*a*) Acceleration at burnout
(*b*) Acceleration at lift-off
(*c*) Maximum altitude

will change as the following are changed, one by one:

1. Combustion chamber pressure, Pc, from 10 to 50 atm
2. Payload, Mpl, from 100 to 2000 kg.
3. Fuel equivalence from 1.25 to 0.75.
4. Nozzle throat area from + 25% to −25%.

The value in doing this problem is twofold: first, to observe what is required to specify a rocket in sufficient detail so that its performance can be analyzed and second, to observe how performance is affected as various features of the rocket are changed.

7.2   For a $C_{10}H_{22}$/liquid oxygen rocket operating at chemically correct propellant ratio, with $Pc = 25$ atm and $Pe = 1$ atm and with a burnout acceleration fixed at 8.5 G, what is the minimum required propellant mass needed to lift a payload of 100 kg to an altitude 750 km?
There are two ways to do the problem:
(*a*) Ignore the mass of structure and engine, and consider the rocket to consist simply of propellant and payload.
(*b*) Assume the required amount of structure and engine are some percentage, say 15%, of the combined mass of payload and propellant.

7.3   Problem 7.2 can be turned around by specifying the mass of propellant and determining the maximum mass of payload that can be carried to some specified altitude.

7.4   For some rocket, the parameters π, β, and Ve are known. Now suppose that another rocket is constructed using the same materials and the same propellants, but with each dimension twice that of the smaller rocket. Suppose further that Ve is the same for both rockets, because the two operate at identical pressures and with the same propellant combination and mixture ratio.
How will the maximum altitudes reached by these two rockets compare? Will they be the same or different, and if different, which will rise to the higher altitude?

7.5   Before attempting problems that involve moon missions, it is a good idea to acquire some appreciation for the variation between burnout velocity at the earth's surface and the time required to travel to the moon.
For $r = 1$, Eq. 7-39 gives the burnout velocity at the earth, and for $r = 59.73$, Eq. 7-39 gives the crash-landing velocity at the moon. Both velocities are influenced, of course, by Vzg, the velocity at the zero-gravity point.
For a series of Vzg values, use Eq. 7-39 to find the corresponding velocities at earth and moon, and use Eq. 7-40 to determine the corresponding flight time by numerical integration.
Plot flight time and moon velocity against earth velocity.

7.6   Before attempting problems that involve moon missions, it is helpful to have a clear understanding of the manner in which rocket velocity varies with altitude during powered flight and during the coasting period after burnout. Equations 7-11 apply during powered flight; the variation of velocity with altitude is easily worked out, assuming that gravitational acceleration is constant at the surface value go.

(*a*) For fixed values of π and Ve, plot velocity against altitude for two or three values of β.
(*b*) For fixed values of β and Ve, plot velocity against altitude for two or three values of π.
(*c*) For fixed values of π and β, plot velocity against altitude for two or three values of Ve.

7.7   The object of this problem is to find (π, β, Ve) combinations that will land a single-stage rocket on the moon. The purpose of the problem is to demonstrate the strong influence that rocket burnout velocity exerts on the time of flight.

If the lift-off and burnout accelerations are to be kept within reasonable bounds, it turns out that all single-stage rockets require nozzle exit velocities Ve that are much larger than can be obtained with present-day propellants. In the next two problems we shall discover that earth-to-moon missions are possible with currently available propellants in multistage rockets.

One way to proceed with the single-stage problem, which is useful by way of showing orders of magnitude, is to first establish the desired flight time from earth to moon. Using the results of Problem 7.5, flight time fixes Vzg, the zero-gravity point velocity, and through Eq. 7-39, we now have the (V, z) curve of rocket velocity, V, as a function of altitude z.

Now choose values for ao and abo, lift-off and burnout accelerations. Together, these define π, and they also define β as a function of Ve, Eqs. 7-22. Through Eqs. 7-12 and 7-13, burnout velocity and altitude, Vbo and zbo, can be found for selected values of Ve. The objective is to find a (Vbo, zbo) combination that falls exactly on the (V, z) curve.

To approximate reality more closely, some tolerance can be specified for the flight time, say, for example, 4 days plus or minus half a day. In this example, then, the 3½ and 4½ flight times established two (V, z) curves, and the area between them defines the "window" into which the rocket must fall at burnout (see Fig. 7.11). It is instructive to discover the small size of the window and the correspondingly small tolerance on maximum and minimum allowable values for Ve. These latter observations explain why mid-course corrections to rocket velocity are required for earth-to-moon missions.

7.8   Repeat Problem 7.7 for a two-stage rocket. A simple way to treat the problem is to assume identical π, β, and Ve values for both stages, so that the accelerations at ignition and burnout are identical. Two variations are possible for the solution:

(a) Assume zero time interval between first-stage burnout and second-stage ignition. In this case, velocity and altitude at second-stage burnout are the sums of the velocity and altitude increments for both stages.

(b) Take into account the finite time interval between first-stage burnout and second-stage ignition. As the delay time lengthens, larger Ve values are required to accomplish the earth-to-moon mission in the same time span.

For either solution, the problem is to discover the required Ve and to compare the Ve values necessary for single- and two-stage rockets.

7.9   Repeat Problem 7.8 for a three-stage rocket. Again, the simple approach is to assume the same π and β values for all three stages. With three stages, the required Ve values begin to fall within the range of nozzle exit velocities attainable with currently available liquid propellants. As in Problem 7.8, the solution can be written with and without delay times between burnout of one stage and ignition of the next.

7.10   In Problems 7.7, 7.8, and 7.9, no mention is made as to what happens at the moon. The interesting case is the soft landing, for which we could specify

Vbo = touchdown velocity at moon

aig = ignition acceleration of the retro-rocket

Assume that the Ve available to the booster rockets is also available for the motor in the retro-rocket.

With (aig, Vbo, Ve) known, the problem now is to discover at what altitude above the moon's surface ignition of the retro-rocket must occur. This can be accomplished by assuming various values for π of the retro-rocket and discovering for which value zig and Vig, the altitude and velocity at ignition, lie on Eq. 7-33 or, better, on Eq. 7-39.

For now, with the π value of the retro-rocket known, we can work backwards and determine the dimensions of the loaded rocket at lift-off on the earth. Do this by assuming a value for the payload to be landed on the moon. The necessary propellant mass and the structural mass plus engine mass required can then be found for the retro-rocket using Eq. 7-41. The total mass of the retro-rocket at ignition is now the payload of the second stage, for which the mass distribution can be determined, and from it we can pass to the first stage for a two-stage booster system.

7.11   With the program developed for Problem 7.10 the characteristics of two- or three-stage rockets can be examined in systematic fashion. The quantities that fix the rocket characteristics are several:

> Number of stages
> Booster burnout acceleration
> Booster ignition acceleration
> Flight time from earth to moon
> Retro ignition acceleration
> Touchdown velocity at the moon
> Mass of payload at the moon
> Delay time between burnout and ignition.

Changing each parameter one by one provides some feel for the factors that have a large influence on the mass of propellant required and the nozzle exit velocity required.

One of the most dramatic illustrations, and one for which common sense and intuition are likely to mislead, is the following: Find the total amount of propellant (i.e., the amount in the retro-rocket and in each booster stage) required to soft-land payloads of

$$100, 1000, \text{ and } 10\ 000 \text{ kg}$$

on the moon. You should discover that a 100-fold increase in ultimate payload does not require anything near that increase in total propellant.

In general, the total propellant required is not particularly sensitive to the configuration of the total rocket system at lift-off. This observation shows us the unavoidable necessity of mid-course corrections to earth-to-moon missions.

7.12   Now turn Problem 7.11 around. Whereas in Problem 7.11, the required Ve is one of the quantities that must be determined, specify a value for Ve. The choice for Ve can be guided by what has been learned in Problem 7.11. If you wish to examine, say, a three-stage rocket, then choose a Ve larger than the required Ve found in Problem 7.11.

We can now specify the character of the rocket system as follows, for example:

Flight time
Maximum booster acceleration at burnout
Minimum booster acceleration at ignition
Ignition acceleration of retro-rocket
Touchdown velocity at the moon
Delay time between burnout and ignition
Payload at moon

With these specifications, the problem is, What is the *minimum* mass of propellant required?

The easy solution assumes that each stage of the booster system has identical π and β values. The solution then consists of a search for the minimum propellant mass, which can be carried out in logical fashion.

There is no guarantee, however, that identical stages is the configuration requiring least propellant. If the individual stages can be designed independently, the solution is vastly more difficult. If you wish to examine solutions of this kind, it is best to start with a two-stage, rather than a three-stage, booster system.

7.13   The nozzle exit velocity Ve occupies a position of central importance in rocket performance. Calculate Ve for a propellant combination of liquid hydrogen/liquid oxygen for

(*a*) various fuel/oxygen ratios
(*b*) various combustion chamber/nozzle exit section pressure ratios, such as 10, 20, 30, and 40.

7.14   Repeat Problem 7.14 for liquid $C_{10}H_{22}$/liquid oxygen.

7.15   Thrust calculations, or thrust per unit area of nozzle throat, require the mass flow rate $\dot{M}$ as well as Ve, since

$$\text{Thrust} = \dot{M}Ve$$

Mass flow rate can be calculated from Eq. 7-28 or by the scheme discussed in connection with Eqs. 7-30a and b. For the conditions of Problem 7.13, calculate the mass flow rate (kg/s/m^2) for liquid hydrogen/liquid oxygen.

7.16   Repeat Problem 7.15 for liquid $C_{10}H_{22}$/liquid oxygen propellant combination.

8
FREE PISTON ENGINES

In a free piston engine, as the name implies, the piston or pistons move in response to forces produced by pressure acting on the surfaces. There is no linkage to a crankshaft, as in conventional prime movers.

Free piston engines are something of a curiosity. Most of the design and development work has been carried out in France, and most of the applications in Europe. As a compressor, or a supplier of high-pressure combustion gas to drive a turbine, the free piston engine has considerable promise.

Engine analysis is straightforward, albeit cumbersome, which may explain in part why these engines are not discussed in introductory texts as standard items.

8.1 DESCRIPTION OF THE ENGINE

Free piston engines can be arranged in various ways to accomplish somewhat different purposes. The essential features of one arrangement are illustrated in Fig. 8.1. The engine is symmetrical and simple. Two pistons oscillate in synchronized, opposed motion between an inner dead point (IDP) and an outer dead point (ODP). Each half of the engine consists of three cylinders, known as the power, compressor, and bounce cylinders.

The working fluid in the power cylinder undergoes a two-stroke Diesel cycle. Fuel injection begins the instant the pistons reach IDP. As burning proceeds, the pistons are driven outward. The intake and exhaust ports in the cylinder wall are uncovered before the piston reaches the ODP. Compression ratios as high as 50:1 can be achieved, so that free piston engines can be operated at a high thermal efficiency and can burn inexpensive, relatively nonvolatile fuel.

The compressor cylinder is equipped with intake and outlet check values. Air is compressed on the inward stroke until the outlet valve opens, after which air is delivered to the power cylinder, or to the surroundings, or to both. The compressor cylinder can be used to supercharge the power cylinder. On the outward stroke of the piston, air intake commences when the pressure in the compressor cylinder drops to ambient.

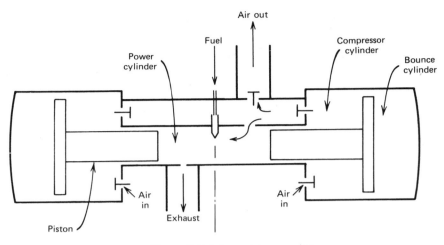

Figure 8.1 A free piston engine.

The bounce cylinders serve as the flywheel, absorbing energy from the piston on the outward stroke and returning it on the inward stroke.

Free piston engines can be constructed to compress air on both strokes, as illustrated in Fig. 8.2. Two-stage compression can be achieved by low-stage compression on the inward stroke, with transfer on the air to the outer compression cylinder, in which second-stage compression takes place on the outward stroke.

Free piston engines are used as gasifiers to drive gas turbines. When used for

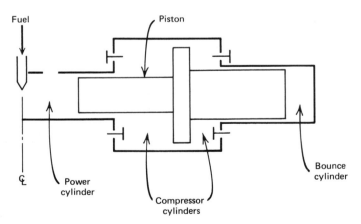

Figure 8.2 A free piston engine, arranged for compression on the inner and outer strokes.

that purpose, the passage marked "Air out" in Fig. 8.1 is blocked off, and all the output from the compressor cylinder passes through the power cylinder and then to the turbine. The exhaust from the power cylinder is diluted and cooled, so that a free piston gasifier delivers high-pressure, low-temperature gases to the turbine. Furthermore, the gasifier operates independently of the turbine rather than being driven by it, and this enhances the flexibility of the combined unit.

The arrangement of opposed pistons virtually eliminates vibration, provided that the pistons are synchronized. To achieve that sort of motion, a mechanical linkage, not shown in Fig. 8.1 or 8.2, is placed between pairs of pistons. Also, the pressures in the bounce cylinders must be identical at all times, which necessitates a pneumatic connection between the two cylinders. Note in Fig. 8.1, where the pistons are shown in approximately the ODP position, that scavenging air from the compressor cylinder enters the power cylinder through the port uncovered by the right-hand piston and leaves the power cylinder through the port opened by the left-hand piston. It is permissible to assume, therefore, that the power cylinder is scavenged completely. This is an important circumstance, and it makes for a substantial simplification in the analysis of engine performance.

8.2 DESCRIPTION OF THE CYCLE

In order to explain the operation of the engine and to determine the motion of the piston, consider the engine shown in Fig. 8.3. By symmetry, only one piston need be considered. Let L denote the length of the power cylinder, the length of the piston, and the combined length of the compressor and bounce cylinders. (The thickness of the piston between the compressor and bounce cylinders is ignored in the interest of reducing the number of parameters required to fix the engine geometry.) The cross-sectional areas of the power and bounce cylinders are denoted by AP and AB,

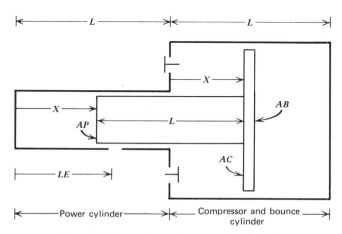

Figure 8.3 Some free piston engine notation.

respectively. The cross-sectional area of the compressor cylinder, AC, is then $AC = AB - AP$. The port location, measured from the power cylinder wall (i.e., the line of symmetry) is LE, and the piston position is identified by X, measured positive to the right, either from the power cylinder wall or from the compressor cylinder wall.

The pressure variations in the three cylinders are illustrated in Fig. 8.4. (The pressure scales are arbitrary.) Here XI and XO denote piston positions at IDP and ODP, respectively.

In the bounce cylinder the pressure varies between PBI and PBO. The pressure-volume relationship will be assumed to be isentropic, since an elementary analysis ignores heat transfer between air and wall, as well as leakage past the piston.

In the compressor cylinder, starting with the piston at XO, air is compressed on the inward stroke. Let PD denote the delivery pressure and PO the ambient pressure. Then XD, the piston position at delivery valve opening, is

$$XD = XO\left(\frac{PO}{PD}\right)^{0.714}$$

8-1

On the outward stroke, compressor cylinder pressure will fall to PO when the piston reaches a position denoted by XC, where

$$XC = XI\left(\frac{PD}{PO}\right)^{0.714}$$

8-2

In the power cylinder, starting with the piston at XO, the pressure remains at PS, the scavenging pressure, until the piston covers the port, after which it increases isentropically to PC, when the piston reaches XI. Fuel is now injected, and the piston moves outward a distance DX during the injection period. After injection ceases, the piston continues to move outward, and the pressure falls isentropically until the port is uncovered. The compression ratio, CR, of the power cylinder is therefore

$$CR = \frac{LE}{XI}$$

8-3

and PC and PS are related by

$$PC = PS\ CR^{1.4}$$

8-4

To start a free piston engine, both pistons are drawn to ODP, and high pressure air from an external source is admitted rapidly to the bounce cylinders. The pistons are driven to IDP, fuel injection begins, the pistons move outward, and the to-and-fro motion of the engine is established.

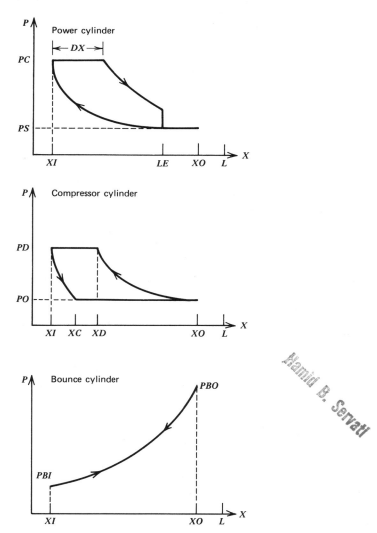

Figure 8.4 Pressure variations in a free piston engine of the type shown in Figs. 8.1 and 8.3.

8.3 KINEMATIC ANALYSIS OF THE CYCLE
Suppose the geometry of an engine of the type drawn in Fig. 8.3 is given:

$$L, AB, AC, AP, \text{ and } LE \text{ are known.}$$

Suppose the operating pressures are given:

$$PO, PS, \text{ and } PD \text{ are known.}$$

The only restrictions on these parameters are

$$AC > 0$$
$$LE < L$$
$$PO \leqslant PS \leqslant PD$$

In a conventional internal combustion engine, the geometry of the crankrod and crankshaft fix the extreme positions of the piston. In a free piston engine, however, the extreme piston positions, XI and XO, are not fixed by geometry but are merely limited by it. Since the engine must pump air on each cycle, the port must be uncovered on the outward stroke, which limits XO,

$$LE < XO < L \qquad\qquad 8\text{-}5$$

and the pressure in the compressor cylinder must reach the delivery pressure PD, which places a limit on XI,

$$0 < XI < XD = XO\left(\frac{PO}{PD}\right)^{0.714} \qquad\qquad 8\text{-}6$$

The key to the kinematics of piston motion derives from the fact that the motion reverses at XI and at XO, and therefore the velocity must be zero at XI and XO. The procedure, then, is to derive expressions for the velocity of the piston as a function of position for the inward and outward strokes. This can be accomplished by deriving expressions for the work done by the gases in the three cylinders on the piston. For the outward stroke, let

$WO(X)$ = the net work done on the piston by the gases in the three cylinders as the piston moves on the outward stroke from XI to X, where $LE < X < L$,

which can be written as

$$WO(X) = OP(X) + OC(X) + OB(X) \qquad\qquad 8\text{-}7$$

where the three terms represent the work done by the gases on the piston for the power, compressor, and bounce cylinders, respectively, on the outward stroke. For the term $OP(X)$, for example,

$$OP(X) = AP \int_{XI}^{X} P_p(X)\, dX, \qquad P_p(X) = \text{pressure in the power cylinder}$$

Referring to Fig. 8.4, the integral can be expanded,

$$OP(X) = AP \left[\int_{XI}^{XI+DX} PC \, dX + \int_{XI+DX}^{LE} PC \left(\frac{XI+DX}{X} \right)^k dX + \int_{LE}^{X} PS \, dX \right]$$

which reduces to

$$OP(X) = AP \, PC \left[DX + \frac{PS}{PC}(X-LE) + \frac{XI+DX}{k-1} \left\{ 1 - \left(\frac{XI+DX}{LE} \right)^{k-1} \right\} \right] \qquad \text{8-8}$$

where k is the heat capacity ratio for the combustion gases in the power cylinder. The two other terms in Eq. 8-7 are

$$OC(X) = AC \, PO \left[X + \frac{XI}{0.4} \frac{PD}{PO} - 3.5 \, XI \left(\frac{PD}{PO} \right)^{0.714} \right] \qquad \text{8-9}$$

$$OB(X) = \frac{PBI \, AB(L-XI)}{0.4} \left[1 - \left(\frac{L-XI}{L-X} \right)^{0.4} \right] \qquad \text{8-10}$$

In similar fashion, let

$WI(X) =$ the net work done on the piston by the gases in the three cylinders as the piston moves on the inward stroke from XO to X, where $0 < X < XD$,

which can be written as

$$WI(X) = IP(X) + IC(X) + IB(X) \qquad \text{8-11}$$

the three terms representing, respectively, the work contributions from the gases in the power, compressor, and bounce cylinders. We have

$$IP(X) = AP \, PS \left[LE - XO + \frac{LE}{0.4} \left\{ 1 - \left(\frac{LE}{X} \right)^{0.4} \right\} \right] \qquad \text{8-12}$$

$$IC(X) = AC \, PO \left[\frac{XO}{0.4} \left\{ 1 - \left(\frac{PD}{PO} \right)^{0.4} \right\} + \frac{PD}{PO} \left\{ X - XO \left(\frac{PO}{PD} \right)^{0.714} \right\} \right] \qquad \text{8-13}$$

$$IB(X) = PBO \, F(X) \qquad \text{8-14}$$

$$F(X) = \frac{AB(L-XO)}{0.4} \left[1 - \left(\frac{L-XO}{L-X} \right)^{0.4} \right] \qquad \text{8-15}$$

Now, since L, LE, AP, AC, and AB are known from the geometry of the engine, and PO, PD, and PS are known (and PC can be obtained from Eq. 8-4), the requirements for zero velocity at XO and XI are

$$WO(XO) = f(XI, XO, PBI, DX, k) = 0 \qquad \text{8-16}$$

$$WI(XI) = g(XI, XO, PBO) = 0 \qquad \text{8-17}$$

In addition, the bounce cylinder pressures at the extreme piston positions, PBI and PBO, are simply related,

$$PBI = PBO\left(\frac{L - XO}{L - XI}\right)^{1.4} \qquad \text{8-18}$$

If we choose a (XI, XO) pair that satisfies the inequalities in Eqs. 8-5 and 8-6, Eq. 8-17 leads at once to the required bounce cylinder pressure at ODP,

$$PBO = -\frac{IP(XI) + IC(XI)}{F(XI)} \qquad \text{8-19}$$

Equation 8-18 provides the value for PBI, and Eq. 8-16 gives the value for DX for an assumed value of k. With the value of DX in hand, the fuel/air ratio in the power cylinder can be computed (see below) and k determined. It is therefore possible to determine a consistent pair of DX and k values by simple iteration. The specific heat ratio, k, for the combustion gases depends on the fuel/air ratio, which will not vary widely. A good approximation is $k = 1.30$.

The solution of Eq. 8-16 for DX can be accomplished by a half-interval search, starting the search in the middle of the allowable range for DX, namely,

$$DX = \frac{LE + XI}{2}$$

Or the solution can be executed with a Newton-Raphson search, for which the recursion formula can be stated as

$$DXNEW = DX - \frac{WO(XO, DX)}{DWO(DX)} \qquad \text{8-20}$$

where DX is the trial value, $DXNEW$ the next approximation, and $DWO(DX)$ is the derivative,

$$DWO(DX) = \frac{d}{d(DX)} WO(XO, DX)$$

$$= \frac{k}{k-1} AP\, PC\left[1 - \left(\frac{XI + DX}{LE}\right)^{k-1}\right] \qquad \text{8-21}$$

In view of the form of the function $WO(XO)$, it is best to choose $DX = 0$ as the first trial value.

To recapitulate, by choosing values for the limits of piston travel, XI and XO, which satisfy the requirement that the engine pump air, Eqs. 8-5 and 8-6, the required bounce cylinder pressure and the required piston movement, DX, during combustion can be determined from Eqs. 8-16 and 8-17. Now the question arises: Can the required DX be realized? If DX is large, there may not be sufficient air in the power cylinder with which to support the combustion of the large mass of fuel required. An approximate criterion with which to answer that question can be formulated in the following way.

Let M_f denote the mass of fuel burned in one cycle; Q_p, the energy released during constant pressure combustion; and M_a, the mass of air in the power cylinder. Then

$$M_f Q_p = M_a c_p\, \Delta T$$

where ΔT is the temperature rise during combustion. The volume change of the power cylinder during combustion is

$$AP\, DX = \frac{M_a R_a\, \Delta T}{PC} \qquad \text{8-22}$$

R_a being the gas constant for air. Eliminate ΔT between these two expressions, and solve for DX,

$$DX = \frac{k-1}{k} \frac{M_f Q_p}{PC\, AP} \qquad \text{8-23}$$

or

$$DX = \frac{k-1}{k} \frac{F}{A} \frac{M_a Q_p}{PC\, AP} \qquad \text{8-24}$$

where F/A is the fuel/air ratio on a mass basis. Since

$$M_a = \frac{PC\, AP\, XI}{R_a TC} \qquad\qquad 8\text{-}25$$

TC denoting the temperature of the air in the power cylinder prior to fuel injection, the expression for DX then takes the form

$$DX = \frac{k-1}{k} \frac{F}{A} \frac{Q_p}{R_a TC} XI \qquad\qquad 8\text{-}26$$

To illustrate the order of magnitude of the proportionality between DX and XI, suppose the pressure ratio in the compression cylinder is 4:1. Then if air enters at 300 K, it will be delivered at

$$300\,(4)^{0.285} = 445 \text{ K}$$

If the compression ratio of the power cylinder is 20:1, the air temperature in the power cylinder prior to combustion, TC, will be

$$TC = 445\,(20)^{0.4} = 1474 \text{ K}$$

For Q_p, take 44 340 kJ/kg as representative for a hydrocarbon. Also, $R_a = 0.287$ kJ/kg K. For air, $k = 1.4$; for combustion products, 1.3. Take the average value, 1.35. From Eq. 8-22, it is clear that DX is proportional to the temperature rise ΔT, and the temperature rise will be largest for a chemically correct mixture, or $F/A = 0.066$ kg/kg for a hydrocarbon. Then the maximum DX which can be achieved for these conditions is

$$DX = \frac{0.35}{1.35}\, 0.066\, \frac{44\,340}{0.287 \times 1475}\, XI = 1.8\, XI$$

To complete the kinematic analysis, if the free piston engine is regarded as an air compressor, then its pumping efficiency is the meaningful figure of merit; we define pumping efficiency as the mass of air delivered to the surroundings per cycle, divided by the mass of fuel burned per cycle. Air is drawn into the compressor cylinder as the piston moves between XC and XO on the outward stroke (see Fig. 8.4). Hence

$$\text{Mass of air } pumped/\text{cycle} = \frac{PO\, AC\,(XO - XC)}{R_a\, TO} \qquad\qquad 8\text{-}27$$

When some of the air *pumped* by the compressor cylinder is diverted to the power cylinder to scavenge the exhaust gases, the mass of air in the power cylinder prior to combustion is to be subtracted from Eq. 8-27 to give the mass of air *delivered* to the surroundings. Using Eq. 8-25, we get

$$\text{Mass air } delivered/\text{cycle} = \frac{PO\,AC(XO - XC)}{R_a\,TO} - \frac{PC\,AP\,XI}{R_a\,TC} \qquad \text{8-28a}$$

The mass of fuel burned during one cycle can be taken from Eq. 8-23

$$\text{Mass fuel burned/cycle} = \frac{k}{k-1}\frac{PC\,AP\,DX}{Q_p} \qquad \text{8-28b}$$

and the delivery efficiency, η_d, can be formulated as

$$\eta_d = \frac{k-1}{k}\frac{Q_p}{R_a\,TC}\left[\frac{TC}{TO}\frac{PO}{PC}\frac{AC}{AP}\frac{XO-XC}{DX} - \frac{XI}{DX}\right] \qquad \text{8-29}$$

Owing to the tangled interdependencies of the various quantities that enter this expression, it is not plainly evident what combination of parameters will produce large efficiency; a complete kinematic analysis of a cycle of operation is required before η_d can be computed.

8.4 DYNAMIC ANALYSIS OF THE CYCLE

The delivery rate of a free piston engine, i.e., the mass of air delivered to the surroundings in unit time, is evaluated from

$$\text{Delivery rate} = \frac{\text{Mass of air delivered/cycle (Eq. 8-28)}}{\text{Period}} \qquad \text{8-30}$$

where the *Period* is the time required to execute a complete cycle. Calculation of the *Period* involves the piston velocity,

$$\text{Period} = \int_{XI}^{XO}\frac{dX}{VO(X)} + \int_{XO}^{XI}\frac{-dX}{VI(X)} \qquad \text{8-31}$$

where $VO(X)$ and $VI(X)$ denote piston velocity as a function of position for outward and inward strokes, respectively. These functions are derived by applying Newton's second law to the piston, which has mass M_p. For the outward stroke, X

$$VO(X)^2 = \frac{2}{M_p}\int_{XI}^{X}[P_p\,AP + P_c\,AC - P_b\,AB]\,dX \qquad \text{8-32}$$

where P_p, P_c, and P_b are the pressures in the power, compressor, and bounce cylinders, respectively. Because of the discontinuities in P_p at $X = XI + DX$ and LE and in P_c at $X = XC$, the integral in Eq. 8-32 must be split into four integrals. Note that this cannot be carried out until the kinematic analysis has been completed and the relative positions of the discontinuities are known. Similarly, for the inward stroke,

$$VI(X)^2 = \frac{2}{M_p} \int_{XO}^{X} [P_b AB - P_c AC - P_p AP] \, (-dX) \qquad 8\text{-}33$$

which must be split into three integrals because of the discontinuities at $X = LE$ and $X = XD$.

Note that when the velocity expressions in Eqs. 8-32 and 8-33 are substituted into Eq. 8-31, the *Period* becomes proportional to $\sqrt{M_p}$. In order to achieve a high delivery rate, we require a short cycle time and therefore a light piston. Piston design is an important feature of free piston engines.

Although straightforward when the pressure-volume variations are assumed to be isentropic, the working out of expressions for $VO(X)$ and $VI(X)$ is a tedious assignment, and is left to the reader.

Since the pressure acting on the piston surfaces differ between the inward and outward strokes, it is not possible to state the errors that would accompany the approximate expressions for the *Period*,

$$\text{Period} \approx 2 \int_{XI}^{XO} \frac{dX}{VO(X)} \approx 2 \int_{XI}^{XO} \frac{dX}{VI(X)}.$$

8.5 PRECISE DETERMINATION OF THE FUEL FLOW RATE

The kinematic analysis provides the value of DX, the length of piston movement during combustion. The analysis given in Eqs. 8-22 through 8-26 produces the approximate expression for the required mass of fuel

$$M_f = \frac{k}{k-1} \frac{PC \, AP \, DX}{Q_p}$$

The exact calculation of the fuel burned in each cycle proceeds as follows: the energy equation for constant pressure adiabatic combustion is

$$N_a h_a(TC) + N_f h_f(TF) = N_p h_p(TP) \qquad 8\text{-}34$$

where N_a, N_f, and N_p denote the mole numbers for air, fuel, and products of combustion; and TC, TF, and TP denote, respectively, air temperature at the start of injection, fuel temperature, and product temperature at the instant injection is completed. The volume change during combustion can be expressed as

$$AP\,DX = \frac{R}{PC}\,(N_p\,TP - N_a\,TC) \qquad\qquad 8\text{-}35$$

These two equations contain three unknowns, N_f, N_p, and TP. But N_a and N_f fix N_p,

$$N_p = f(N_a, N_f) \qquad\qquad 8\text{-}36$$

by the rules for complete combustion. The mole number for the fuel, N_f, can now be determined by a trial-and-error search and the fuel mass, M_f, evaluated from the known molecular weight of the fuel.

8.6 SUMMARY

A free piston engine is a relatively simple machine for which the kinematic and dynamic operating characteristics are obtained from a fairly complex algebraic analysis. The interesting feature of the engine is the fact that the length of the piston stroke, $XO - XI$, can be varied, within limits, for a given engine. Furthermore, for a given stroke the position of stroke in the cylinder can be varied, within limits. This flexibility is not characteristic of conventional prime movers.

Figure 8.5 illustrates a few performance characteristics of a simple free piston compressor of the type drawn in Fig. 8.3. The engine geometry is fixed, the ambient and delivery pressures are fixed, and the piston position at the outer dead point is varied, keeping the piston position at the inner dead point fixed. Essentially, Fig. 8.5 illustrates how the pumping efficiency of the compressor is affected by the length of the piston stroke. High efficiency calls for a short stroke. Note the huge rise in PBO, the pressure in the bounce cylinder with the piston in the outer dead position, when the piston stroke approaches the maximum allowable value.

Figure 8.6 shows how the piston velocity varies with piston position. For a good part of the stroke, the velocities during outward and inward motion are identical. As the piston approaches the inner dead position; the inward velocity drops below the outward velocity as a result of the decelerating effect of compressing air in the power cylinder. For other combinations of engine parameters, the difference between inward and outward velocities can be substantially larger than those shown in Fig. 8.6.

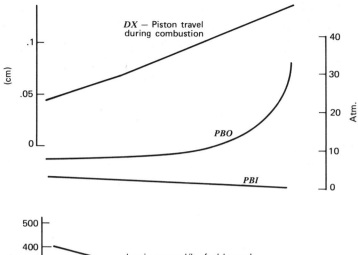

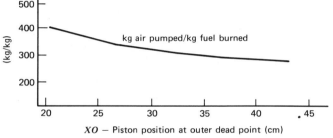

Figure 8.5 Free piston compressor performance.

L = 45.7 cm	AB = 452 cm^2	Q_p = 44 300 kJ/kg fuel
LE = 20.3 cm	AP = 323 cm^2	k = 1.36 (assumed)
XI = 1.0 cm	PO = 1 atm	
CR = 20:1	PS = 1 atm	
	PD = 4 atm	

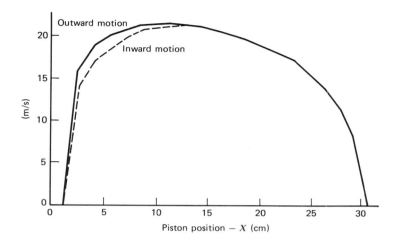

Figure 8.6 Piston velocity and piston position.

L = 45.7 cm	AB = 729 cm^2	Q_p = 44 300 kJ/kg fuel
LE = 25.4 cm	AP = 365 cm^2	k = 1.36 (assumed)
XI = 1.27 cm	PO = 1 atm	Piston mass = 22.2 kg
XO = 30.5 cm	PS = 1 atm	
CR = 20:1	PD = 6.8 atm	

PROBLEMS

8.1 The purpose of this problem is to examine the effect of various engine parameters on pumping efficiency, as defined in Eq. 8.29.

For an engine of the type illustrated in Fig. 8.3, let

$$Q_p = 45\ 000 \text{ kJ/kg}$$
$$k = 1.3$$
$$R_a = .287 \text{ kJ/kg K}$$
$$PO = 1 \text{ atm}$$
$$TO = 300 \text{ K}$$

The following design details of the engine,

$$L, LE, AP, \text{ and } AB$$

and the following operating variables,

$$PD, PS, XI, \text{ and } XO$$

may each be varied independently, within limits, of course. Note how each affects pumping efficiency. Note also that the quantity $XO - XI$ can remain constant while XO and XI are varied.

    8.2   Continuing with Problem 8.1, examine the manner in which engine dimensions and operating conditions affect the pumping rate, Eq. 8-30. The time occupied during one cycle of the engine is needed, which introduces the mass of the piston. Piston mass will depend on the material from which it is constructed, together with its shape. Figure 8.7 suggests the obvious shape. The simple approach would maintain the thickness dimension t constant, independent of the forces acting on the piston. A more sophisticated treatment would adjust thickness to limit the amount of bending.

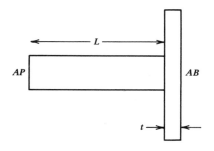

Figure 8.7

    8.3   The simple shape of a free piston engine of the type illustrated in Fig. 8.3 suggests the following problems.

    Suppose the diameter and length of a cylindrical space are defined. What engine design and operating conditions will produce maximum delivery of air?

    Suppose the magnitude, but not the shape, of a cylindrical space is defined. What engine design and operating conditions will produce maximum delivery of air?

    Assume atmospheric temperature and pressure at air intake port and a fixed air delivery pressure, *PD*. Also, maximum allowable pressure in the power cylinder and in the bounce cylinder should be specified. The choice of piston material is arbitrary, so that the search for optimum engine configuration does not include piston material as an assignable variable.

9
BOILERS AND FURNACES

9.1 INTRODUCTION

Boilers burn fuel and air in order to convert a steady flow of liquid water into steam. Furnaces generally produce higher temperatures than do boilers, in order to heat-treat nuts and bolts moving continuously along a conveyor belt, for example. The objective of this chapter is to examine the methods for calculating the rate of heat loss from a boiler or furnace for three cases: oil-fired boiler, coal-fired boiler, and wood-fired boiler. The methods in the three cases are in some respects similar and in other respects are not.

Figure 9.1 shows the essential features of a boiler. When we assume steady flow, the energy equation is

$$\dot{m}_f h_f(t_f) + \dot{m}_a h_a(t_a) + \dot{m}_w h_w(1) = (\dot{m}_a + \dot{m}_f)h_p(t_p) + \dot{m}_w h_w(2) + \dot{Q} \qquad 9\text{-}1$$

Subscripts a, f, p, and w denote air, fuel, products of combustion, and water, respectively. The $\dot{m}$ denotes mass flow rate, and $\dot{Q}$ is the rate of heat loss, taken positive from the boiler to the surroundings.

Solving for $\dot{Q}$

$$\dot{Q} = \dot{m}_f[h_r(t_r) - h_p(t_p)] + \dot{m}_w[h_w(p_1, t_1) - h_w(p_2, t_2)] \qquad 9\text{-}2$$

Here, subscript r denotes the reactants, and h_r and h_p must be evaluated per unit mass of fuel.

Suppose, now, that the four temperatures t_r, t_p, t_1 and t_2, and the two pressures p_1 and p_2 have been measured. In addition, the water flow rate, $\dot{m}_w$, has been measured, and the chemical composition of the fuel is known. The problem of calculating $\dot{Q}$ can be cast in three forms:

1. We know $\dot{m}_a$ and $\dot{m}_f$ and the composition of the product gases in the exhaust stack as they leave the combustion chamber. With this information, the best estimate of $\dot{Q}$ will result.

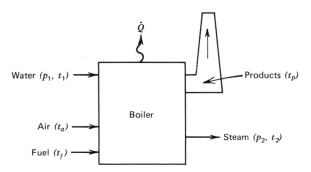

Figure 9.1 A simple boiler.

2. We know $\dot{m}_a$ and $\dot{m}_f$. In this case, we must make a guess at the composition of the product gas mixture in order to evaluate the term $h_p(t_p)$.
3. We know $\dot{m}_f$ and the composition of the product gas mixture. We must now estimate the air flow rate, $\dot{m}_a$, in order to assign a numerical value to the term $h_r(t_r)$.

We will examine the last case. The analysis reveals how we could proceed in the other two cases.

9.2 OIL-FIRED BOILER

The rate of heat loss is to be calculated for the following set of operating conditions:

$$\text{Fuel} - \text{liquid propane, } C_3H_8$$
$$\dot{m}_f = 45.4 \text{ kg/hr}$$
$$\dot{m}_w = 545 \text{ kg/hr}$$
$$t_r = t_a = t_f = 27°C$$
$$t_p = 327°C$$
$$t_1 = 20°C, p_1 = 5 \text{ atm}$$
$$t_2 = 150°C, p_2 = 3 \text{ atm}$$

and the flue gas analysis is reported as follows:

Gas	Percent by volume
N_2	84
CO_2	8
CO	3
O_2	5
	100

Since the combustion of propane must produce H_2O vapor in the products, the above is a "dry" analysis, that is, the water vapor was condensed out prior to the chemical analysis. To make use of the analysis, recall that a statement about the composition of a mixture of ideal gases on a volume basis is at once a statement of the composition on a mole basis. This means that 84% of the particles in the *dried* products are N_2 molecules, 8% CO_2 molecules, and so forth. Ignoring the water vapor that may be present in the air (see the next case, coal-fired boiler, which will include incoming water vapor), we can estimate the amount of water vapor in the products in two ways. The fuel and products must contain the same H/C atom ratio, hence

$$\left(\frac{H}{C}\right)_{fuel} = \frac{2n_{H_2O}}{n_{CO} + n_{CO_2}} \qquad 9\text{-}3$$

where n is the mole number. In the present case

$$n_{H_2O} = \frac{1}{2}\frac{8}{3}(8 + 3) = 14.7 \text{ mol}$$

On the other hand, since there is no oxygen in the fuel, the product mixture should have the same N/O ratio as air, and therefore

$$\left(\frac{N}{O}\right)_{air} = \frac{2n_{N_2}}{n_{H_2O} + n_{CO} + 2(n_{CO_2} + n_{O_2})} \qquad 9\text{-}4$$

which leads to a second value for n_{H_2O},

$$n_{H_2O} = \frac{2 \times 84}{3.76} - 3 - 2(8 + 5) = 15.7 \text{ mol}$$

The discrepency means that the gas analysis is crude. Which value should be used? The first value, derived from the C/H value for the fuel, is preferred. When the second value is used, the propane/air reaction will not produce the required amount of H_2O in the products, which is awkward. We therefore take as the product mixture

$$8CO_2 + 3CO + 5O_2 + 14.7H_2O + 84N_2$$

That mixture contains 11 mol of carbon. A mole of fuel contains 3 mol of carbon. When we multiply by the factor 3/11, the product mixture is

$$2.2CO_2 + 0.8CO + 1.4O_2 + 4.0H_2O + 22.9N_2 \qquad 9\text{-}5$$

from which the reactant mixture is derived,

$$C_3H_8 + 6.0O_2 + 22.9N_2 \qquad\qquad 9\text{-}6$$

The reactant mixture contains a mole of fuel, which is convenient but in no way necessary. We can now evaluate the first term on the right-hand side of Eq. 9-2

$$h_r(t_r) - h_p(t_p)$$

From the definition for the heat of reaction of liquid propane we can find the enthalpy of propane and with it evaluate the term $h_r(t_r)$, knowing the composition from expression 9-6. The chemically correct fuel/oxygen reaction for propane is

$$C_3H_8 + 5O_2 \rightarrow 3CO_2 + 4H_2O$$

Therefore

$$H_{rp} = (3h_{CO_2} + 4h_{H_2O} - h_{C_3H_8} - 5h_{O_2})_{298}$$

From Table E.4, Appendix E,

$$H_{rp} = -2\,016\,900 \text{ kJ/kmol propane}$$

Solving for the fuel term, and using enthalpy values from Table E.2, Appendix E

$$h_{C_3H_8} = 2\,188\,000 \text{ kJ/kmol}$$

Then

$$h_r(300) = (h_{C_3H_8} + 6.0h_{O_2} + 22.9h_{N_2})_{300}$$
$$= 2\,653\,000 \text{ kJ/kmol propane}$$

For the product term, using expression 9-5,

$$h_p(600) = (2.2h_{CO_2} + 0.8h_{CO} + 1.4h_{O_2} + 4.0h_{H_2O} + 22.9h_{N_2})_{600}$$
$$= 1\,155\,000 \text{ kJ/kmol propane}$$

hence

$$h_r(t_r) - h_p(t_p) = 1\,500\,000 \text{ kJ/kmol propane}$$

For the second term on the right-hand side of Eq. 9-2, the water enters the boiler as a compressed (or subcooled) liquid. Since the pressure is not large, 5 atm, take the entering enthalpy as saturated liquid:

$$h_w(20°C, 5 \text{ atm}) = h_w(20°C, \text{ sat liq})$$
$$= 84 \text{ kJ/kg}$$
$$h_w(150°C, 3 \text{ atm}) = 2761 \text{ kJ/kg}$$

and

$$\dot{m}_w[h_w(p_1, t_1) - h_w(p_2, t_2)] = 1\ 459\ 000 \text{ kJ/hr}$$

For $\dot{m}_f$,

$$\dot{m}_f = 45.4 \text{ kg/hr}$$
$$= 1.03 \text{ kmol/hr}$$

so that

$$\dot{m}_f[h_r(t_r) - h_p(t_p)] = 1\ 551\ 000 \text{ kJ/hr}$$

giving a rate of heat loss

$$\dot{Q} = 1\ 551\ 000 - 1\ 459\ 000$$
$$= 92\ 000 \text{ kJ/hr}$$

The solution for the rate of heat loss employs the measured composition of the product gases as information that leads to the composition of the reactant mixture; introducing the thermodynamic definition of the heat of reaction makes it possible to evaluate the enthalpy difference $[h_r(t_r) - h_p(t_p)]$.

9.3 COAL-FIRED BOILER

For a coal-fired boiler, the rate of heat loss may be computed by the same procedure outlined in the previous section. Suppose the air and coal enter the combustion chamber at 300 K, and suppose the product gas mixture enters the exhaust stack at 600 K. We will evaluate the term

$$h_r(300) - h_p(600)$$

The coal is anthracite, with the following composition:

	Percent by weight	Atomic weight	Moles
C	84.5	12	7.0
H	3.5	1	3.5
O	4.2	16	0.26
S	0.9	32	0.03
N	1.5	14	0.11
Ash	5.4	60	0.09
	100		

The sample had been dried before analysis. Coal "as received" contains liquid water. For the ash, we could assume it to be SiO_2, with a molecular weight of 60. Dividing the weight fractions by the atomic weights produces coal composition on a mole basis. In what follows, the sulfur and ash will be ignored, since we lack thermodynamic data on both. For anthracite coal, this will not introduce appreciable error. Such would not be the case for a high-sulfur coal.

Suppose the dry analysis of the flue gas mixture is found as follows:

Gas	Percent by Volume
CO_2	11
CO	5
N_2	79
O_2	5
	100

The amount of water vapor in the products is estimated from the (H/C) ratio of the coal, using Eq. 9–3

$$n_{H_2O} = \frac{1}{2} \frac{3.5}{7.0} (11 + 5) = 4.0$$

The product gas mixture can then be set down approximately (because the ash and sulfur are missing) as

$$11CO_2 + 5CO + 5O_2 + 4H_2O + 79N_2$$

where the coefficients are moles. Reducing these coefficients by the ratio 7:16 results in a product mixture

$$4.8CO_2 + 2.2CO + 2.2O_2 + 1.8H_2O + 34.6N_2 \qquad 9\text{-}7$$

and if these coefficients are in units of gram moles (gmoles), that product mixture is obtained by burning 100 g of coal. The reactant mixture is

$$(7.0C + 1.8H_2 + 0.13O_2 + 0.05N_2) + 9.0O_2 + 34.5N_2 \qquad 9\text{-}8$$

where the four terms in brackets represent 100 g of coal.

The measured heat of reaction for this coal sample is reported as $-14\,030$ Btu/lbm dry coal. Had that calorimetric determination not been available, the heat of reaction could be calculated from Dulong's empirical equation

$$U_{rp} = 14\,544C + 62\,028\left(H - \frac{O}{8}\right) + 4050S$$

where C, H, O, and S are percentages by weight. In the present case, Dulong's equation leads to

$$U_{rp} = 12\,290 + 1845 + 36 = 14\,171 \text{ Btu/lbm dry coal}$$

understood to be negative. Note the insignificant contribution from the sulfur term. Ignoring the difference between H_{rp} and U_{rp}, we have

$$H_{rp} = -14\,030 \text{ Btu/lbm dry coal}$$
$$= -3250 \text{ kJ/100 g dry coal}$$

For this anthracite coal, the chemically correct coal/oxygen reaction is

$$(7.0C + 1.8H_2 + 0.13O_2) + 7.8O_2 \rightarrow 7CO_2 + 1.8H_2O$$

and the thermodynamic statement for the heat of reaction is

$$H_{rp} = (7.0h_{CO_2} + 1.8h_{H_2O} - h_{coal} - 7.8h_{O_2})_{300}$$

from which the h_{coal} term can be secured

$$h_{coal}(300) = (7h_{CO_2} + 1.8h_{H_2O} - 7.8h_{O_2})_{300} + 3250$$
$$= 3280 \text{ kJ/100 g dry coal.}$$

Then $h_r(300)$ for the reactant mixture in expression 9-8 is

$$h_r(300) = (h_{coal} + 9.0h_{O_2} + 34.5h_{N_2})_{300}$$
$$= 4000 \text{ kJ}/100 \text{ g dry coal}$$

and $h_p(600)$ for the product mixture in expression 9-7 is

$$h_p(600) = (4.8h_{CO_2} + 2.2h_{CO} + 2.2h_{O_2} + 1.8h_{H_2O} \ 34.6h_{N_2})_{600}$$
$$= 1800 \text{ kJ}/100 \text{ g dry coal}$$

and therefore

$$h_r(300) - h_p(600) = 2200 \text{ kJ}/100 \text{ g dry coal}$$

If the air had been saturated with water vapor at $77°F$ and the coal contained 6% water by weight, then terms must be added to both the reactant and product mixtures. Each 100 g of dry coal is accompanied by 6.4 g or 0.35 gmol of liquid water. For saturated air at $77°F$, the specific humidity is 140 grains of water vapor/lbm of dry air. (7000 grains = 1 lbm) The specific humidity is then

$$\frac{140}{7000} \times \frac{28.95(\text{g/gmol air})}{18(\text{g/gmol H}_2\text{O})} = 0.032 \ \frac{\text{gmol H}_2\text{O}}{\text{gmol air}}$$

The reactant mixture in expression 9-8 contains 100 g of dry coal and

$$9 + 34.5 = 43.5 \text{ gmol of dry air}$$

The mass of water vapor in the air is then

$$0.032 \ \frac{\text{gmol H}_2\text{O}}{\text{gmol air}} \times 43.5 \ \frac{\text{gmol air}}{100 \text{ g dry coal}} = 1.4 \ \frac{\text{gmol H}_2\text{O}}{100 \text{ g dry coal}}$$

The term $h_r(300)$ is now

$$h_r(300) = 4000 + 0.35h_{H_2O}(\text{liq}) + 1.4h_{H_2O}(\text{gas})$$

and approximately,

$$h_{H_2O}(\text{liq}) = h_{H_2O} - h_{fg}$$

where h_{fg} is the heat of vaporization at 300 K. Then

$$h_{H_2O}(\text{liq}) = 57\ 376 - 44\ 000$$
$$= 13\ 420\ \text{kJ/gmole}$$

Now

$$h_r(300) = 4000 + 0.35 \times 13\ 420 + 1.4 \times 57\ 376$$
$$= 4085\ \text{kJ/100 g dry coal}$$

and

$$h_p(600) = 1800 + (0.35 + 1.4)h_{H_2O}$$
$$= 1900\ \text{kJ/100 g dry coal}$$

and the enthalpy difference is now

$$h_r(300) - h_p(600) = 2185\ \text{kJ/100 g dry coal}$$
$$= 2185\ \text{kJ/106 g coal}$$
$$= 2060\ \text{kJ/100 g coal.}$$

The comparison between dry coal and dry air, on the one hand, and wet coal and saturated air, on the other, is between 2200 and 2185 kJ/100 g dry coal, a difference of less than 2%.

With $[h_r(300) - h_p(600)]$ known, the rate of heat loss may be calculated from Eq. 9-2 for the given flow rates and water inlet and outlet conditions.

9.4 WOOD-FIRED BOILER

Wood fiber can be represented satisfactorily by $C_6H_{10}O_5$, but the heat of reaction will vary between 18 000 and 36 000 kJ/kg of fiber, depending on the character and amount of resin present. Furthermore, green wood may contain as much as 50% moisture by weight; seasoned wood generally shows a moisture content of between 10 and 25%.

Without the benefits of a calorimetric measurement of the heat of reaction, any and all thermodynamic calculations that have to do with wood combustion may be incorrect by a factor of at least 2.

9.5 SUMMARY

Analysis of rate of heat loss from boilers and furnaces is based on a steady flow energy equation, which will contain a pair of terms

$$h_r(T_r) - h_p(T_p)$$

where r and p represent reactants and products, respectively, and the enthalpies must both be in units per mass of fuel. The term h_r will include the enthalpy of the fuel, which can be assigned a numerical value by introducing the thermodynamic definition of the heat of reaction.

The composition of the reactant mixture is either known from measurements of fuel and air flow or is calculated from the measured composition of the product gases. This latter calculation, when required, is generally the weakest link in the total computation and limits the accuracy we can expect to achieve.

10
CHEMICAL EQUILIBRIUM

10.1 INTRODUCTION

In the preceding chapters, the temperature reached following combustion was determined by applying two general principles, conservation of energy and matter, together with certain assumptions about the overall chemistry of the combustion reaction. Those assumptions, referred to as complete combustion, made it possible to write down the composition of the product gas mixture merely by inspecting the reactant mixture. Thus a hydrocarbon/air mixture when burned will produce

$$CO_2, H_2O, O_2, \text{ and } N_2$$

if there is excess air, and

$$CO, CO_2, H_2O, \text{ and } N_2$$

if there is excess fuel. Under certain conditions, these gases are an accurate representation of the product mixture; under other conditions, they are not. In this chapter it will be shown how these simplifying assumptions can be set aside and replaced by the requirements of the second law.

10.2 ELEMENTARY KINETIC THEORY

One of the early triumphs of the kinetic theory was to show that the average velocity of gas particles is simply related to gross properties of the gas. The demonstration proceeds from the assumption that a gas exerts pressure as the result of momentum change during impact of a gas particle with the wall of the confining vessel. For a system of N particles, each with mass m and considered to be hard, elastic spheres moving with velocity c, the product of pressure P and vessel volume V is simply proportional to the total kinetic energy

$$PV = \frac{2}{3} \left(\frac{1}{2} mNc^2 \right)$$

Consequently, for a gas, pressure is proportional to kinetic energy per unit volume. In addition, of course, it was discovered by experiment that

$$PV = MRT$$

where M is total mass, R is a constant whose value depends on the gas in question, and T is the absolute temperature or, more simply, a function of ordinary temperature that makes the equation valid. It follows at once that the average velocity c is related to T and molecular weight, mw,

$$c \sim \sqrt{\frac{T}{mw}}$$

At room temperature, oxygen molecules move at about 450 m/s, and hydrogen molecules move at 1800 m/s. These are the average molecular velocities. As the result of collisions, some molecules have velocities that are higher than the average, and some are lower. The velocity of any particular molecule changes rapidly with time because of the high collision frequency. At room temperature and pressure, collisions between pairs of hydrogen molecules occur at the rate of about one billion per second.

As the temperature of a gas composed of hard elastic spheres increases, the average particle velocity increases. When a real gas, such as hydrogen, is heated, the situation is more complex. A pair of hydrogen atoms are held in a molecular configuration under the action of an interatomic force system that has the general character shown in Fig. 10.1. The atoms repel on close approach and attract when drawn far apart. For large separation, the force rapidly attenuates to zero.

When hydrogen is heated, the added energy manifests itself in the mechanical motion of the atoms and molecules. For a diatomic molecule, three forms of motion

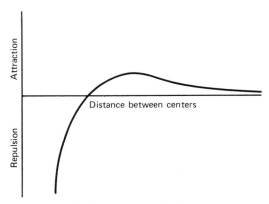

Figure 10.1 An inter-atomic force system.

are possible: translation, rotation, and vibration. Translation refers to the movement of the center of gravity of the molecule. The atoms may rotate about an axis perpendicular to the line joining their centers; also, that axis may rotate. Vibration refers to the to-and-fro motion along the line joining the centers of the two atoms.

One of the basic theorems of kinetic theory is the notion of equipartition of energy; there is no preferred form of motion. The energy in a gas is distributed among all the forms of motion, and what is more, the distribution is even when summed over the whole array of particles. When a diatomic gas is heated, the translational velocities must increase, and the rotation and vibration must become more vigorous. If the two latter forms of motion become too energetic in some molecules, the weak attractive force between atoms will be exceeded, and dissociation of those molecules results, so for hydrogen,

$$H_2 \longrightarrow 2H$$

The hydrogen atoms produced by dissociation acquire a range of velocities, by collision. If a pair collide under proper conditions they will associate and form a molecule,

$$2H \longrightarrow H_2$$

When hydrogen is heated, and the level of activity of all forms of motion is increased, there will be more molecules in a condition for dissociation than there will be pairs of atoms associating. With an increase in temperature, the amount of molecular hydrogen present will decrease, while the amount of atomic hydrogen will increase. When the temperature decreases, the effects will be oppositely directed. Because of the high collision rate, a steady-state composition is reached quickly.

The conditions required for steady-state composition in a diatomic gas system can be formulated readily. The number of molecules per unit volume is proportional to the partial pressure, p_m. Only a fraction, f_m, of the molecules will become hyperactive through collisions in the next instant. The number of dissociation reactions occurring in the next instant is proportional to

$$p_m f_m$$

The number of atoms per unit volume is proportional to the partial pressure exerted by the atoms, p_a. The number of atomic collisions occurring in the next instant is proportional to p_a^2. Of all atomic collisions, only a fraction, f_a, will result in the formation of a molecule. For steady-state conditions, we must have

$$p_m f_m = f_a p_a^2 \qquad\qquad 10\text{-}1$$

Presently, the task of assigning values to f_m and f_a, based on the elementary properties of the molecules and atoms, is beyond our reach. However, thermodynamics provides a means for deriving the counterpart to Eq. 10-1, based on the second law.

10.3 THE LAW OF MASS ACTION

Consider the arrangement illustrated in Fig. 10.2. A rigid, insulated vessel contains five gases, A_1, A_2, A_3, A_4, and A_x. The number of moles of each gas are represented by n_i. Gas A_x is chemically inert, but the remaining four gases can interact through the reversible reaction

$$\nu_1 A_1 + \nu_2 A_2 \;\rightleftharpoons\; \nu_3 A_3 + \nu_4 A_4 \qquad\qquad 10\text{-}2$$

where the ν_i are stoichiometric coefficients. The pressure and temperature in the mixture are P and T.

We want to be able to answer the question: Is the system of gases in chemical equilibrium? First, what do we mean by the question? When the system is in thermodynamic equilibrium, we mean that none of its thermodynamic parameters will change with time. If the system is in chemical equilibrium, we mean that none of the thermodynamic parameters, and none of the chemical parameters, will change with time. In the present case, the system will be in chemical equilibrium if there is no tendency for $n_1, n_2, n_3, n_4, n_x, P$, and T to change.

Chemical equilibrium means thermodynamic equilibrium and the absence of a tendency for a reversible chemical reaction to proceed in one direction in preference to the other direction.

In order to answer the question of chemical equilibrium, note that the system is isolated. The rigid, insulated vessel precludes the possibility of energy exchanges between the system and its surroundings, either by heat or work transfers. According to the second law, in an isolated system the entropy can only increase. If the entropy of a system is stationary for all allowable variations in the parameters of the system, equilibrium prevails. If the entropy can be changed by the variations of one or more parameters, equilibrium has not been reached. The procedure, then, is to formulate the

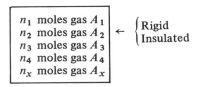

Figure 10.2 An isolated system containing a mixture of reactive gases.

total system entropy and apply the constraints. For a mixture of ideal gases, the total entropy, S, is

$$S = \sum_i n_i s_i$$

where the summation extends over all the gases present, active and inert, and s_i denotes a partial molal entropy. Then

$$dS = \sum_i n_i \, ds_i + \sum_i s_i \, dn_i \qquad \text{10-3}$$

The entropy differentials may be eliminated by substituting

$$ds_i = \frac{du_i + p_i \, dv_i}{T} \qquad \text{10-4}$$

where T is not subscripted because temperature equilibrium is assumed. (Note that the use of Eq. 10-4, the first law, assumes that the system is in equilibrium. Classical thermodynamics is restricted to discussions of equilibrium states.) The internal energy of the gas system, U, is constant,

$$U = \sum_i n_i u_i = \text{constant}$$

so that

$$\sum_i n_i \, du_i = - \sum_i u_i \, dn_i \qquad \text{10-5}$$

with which the du_i terms may be eliminated from the expression for dS. Since the volume is constant,

$$V = n_i v_i = \text{constant}, \qquad \text{for all } i$$

so that

$$dv_i = - \frac{v_i \, dn_i}{n_i}, \qquad \text{for all } i \qquad \text{10-6}$$

and the dv_i terms may be eliminated from the expression for dS. Making use of the definition of enthalpy,

$$h = u + pv$$

the differential of the total entropy has the form

$$T\,dS = -\sum_i (h_i - Ts_i)\,dn_i \qquad\qquad 10\text{-}7$$

Now the mole numbers cannot change independently. Since A_x is an inert gas, $dn_x = 0$. Furthermore, conservation of matter requires that

$$\frac{dn_1}{\nu_1} = \frac{dn_2}{\nu_2} = -\frac{dn_3}{\nu_3} = -\frac{dn_4}{\nu_4} \qquad\qquad 10\text{-}8$$

When we use Eqs. 10-8 to eliminate dn_2, dn_3, and dn_4 from Eq. 10-7, the result is

$$T\,dS = [\nu_3(h_3 - Ts_3) + \nu_4(h_4 - Ts_4) - \nu_1(h_1 - Ts_1) - \nu_2(h_2 - Ts_2)]\,\frac{dn_1}{\nu_1} \qquad 10\text{-}9$$

Equation 10-9 relates the change in S to the change in n_1. If n_1 does not or cannot change, then S cannot change. That situation, while certainly true, is neither interesting nor informative. On the other hand, if the term in brackets in Eq. 10-9 is zero, that is if

$$\nu_1(h_1 - Ts_1) + \nu_2(h_2 - Ts_2) = \nu_3(h_3 - Ts_3) + \nu_4(h_4 - Ts_4) \qquad 10\text{-}10$$

then $dS = 0$ for *all* values of dn_1, positive, negative, or zero. As time passes, n_1, as well as n_2, n_3, and n_4, changes, through the reaction 10-2. As long as Eq. 10-10 remains valid, S will not change, and Eq. 10-10 is therefore a statement of the conditions under which the system is in chemical equilibrium. However, Eq. 10-10 is not a useful form. When the partial molal entropy is expanded (see Appendix A)

$$s_i = \int_0^T c_{pi}\,\frac{dT}{T} - R\,\ln(p_i) + s_{oi}$$

where s_{oi} denotes the datum value at zero absolute temperature and 1 atm pressure, the term $(h_i - Ts_i)$ takes the form

$$h_i - Ts_i = RT\,\ln(p_i) + f_i(T)$$

with   10-11

$$f_i(T) = h_i - T\int_0^T c_{pi}\,\frac{dT}{T} - Ts_{oi}$$

and in place of Eq. 10-10, we have a more concise expression, namely,

$$K_p = \frac{p_3^{\nu_3} p_4^{\nu_4}}{p_1^{\nu_1} p_2^{\nu_2}}$$

10-12

Equation 10-12 is referred to as the *law of mass action*. It is a relation between the *reaction constant*, K_p, and the partial pressures and stoichiometric coefficients of the *reactive gases*. While Eq. 10-12 does not contain any quantity that relates to the inert gas A_x, the influence of that gas is nevertheless present because the pressures of the reactive gases' are partial pressures.

When Eq. 10-12 is satisfied, the system of five gases is in chemical equilibrium; otherwise it is not.

The expression for the reaction constant K_p is, using the f_i function defined in Eq. 10-11,

$$\ln(K_p) = \frac{\nu_1 f_1 + \nu_2 f_2 - \nu_3 f_3 - \nu_4 f_4}{RT}$$

10-13

which can be expanded to read

$$\ln(K_p) = \frac{\nu_1 h_1 + \nu_2 h_2 - \nu_3 h_3 - \nu_4 h_4}{RT}$$
$$- \frac{1}{R} \int_0^T \frac{\nu_1 c_{p1} + \nu_2 c_{p2} - \nu_3 c_{p3} - \nu_4 c_{p4}}{T} \, dT$$
$$+ R(\nu_1 s_{o1} + \nu_2 s_{o2} - \nu_3 s_{o3} - \nu_4 s_{o4})$$

10-13a

The s_{oi} terms are constants. For ideal gases, c_p and h are functions of temperature only. For real gases, they are both functions of temperature and very weak functions of pressure. Consequently, we can regard K_p as *a function of temperature only*.

Differentiating Eq. 10-13a with respect to T, we write it in three lines, each corresponding to the same line in 10-13a, thus

$$\frac{d}{dT} \ln(K_p) = \frac{\Delta H}{RT^2} + \frac{\nu_1 c_{p1} + \nu_2 c_{p2} - \nu_3 c_{p3} - \nu_4 c_{p4}}{RT}$$
$$- \frac{\nu_1 c_{p1} + \nu_2 c_{p2} - \nu_3 c_{p3} - \nu_4 c_{p4}}{RT}$$
$$+ 0$$

10-13b

which will reduce to a neat and concise form

$$\boxed{\frac{dK_p}{K_p} = \frac{\Delta H}{RT^2} dT}$$

10-14

This important and useful equation, known as *van't Hoff's equation*, relates the reaction constant, K_p, and the heat of reaction, ΔH, for reaction 10-2,

$$\Delta H = \nu_3 h_3 + \nu_4 h_4 - \nu_1 h_1 - \nu_1 h_1$$

10-15

Reaction 10-2 is a reversible reaction. When A_3 and A_4 appear on the right-hand side, we choose to identify them as the products, and they enter Eq. 10-15 as positive quantities. Prior to quantum mechanics the functional relationship between K_p and T was secured by first measuring the equilibrium chemical composition in the laboratory at some convenient temperature and pressure. Using the value of K_p computed from Eq. 10-12 for that measured composition as the integration constant, van't Hoff's equation is integrated. Table C.2, Appendix C, lists the a, b, c, d coefficients for various reversible reactions that can be used in the K_p expression,

$$K_p = \exp\left[\frac{a}{T} + \left(b + \frac{c}{T}\right)\ln(T) + d\right]$$

10-16

By introducing the total pressure and total mole number, P and N, respectively, two alternate forms of the mass action law can be written

$$K_p = \frac{n_3^{\nu_3} n_4^{\nu_4}}{n_1^{\nu_1} n_2^{\nu_2}} \left(\frac{P}{N}\right)^{\nu_3 + \nu_4 - \nu_1 - \nu_2}$$

10-17

$$K_p = \frac{x_3^{\nu_3} x_4^{\nu_4}}{x_1^{\nu_1} x_2^{\nu_2}} (P)^{\nu_3 + \nu_4 - \nu_1 - \nu_2}$$

10-18

where x_i is the mole fraction of gas i.

In Eqs. 10-17 and 10-18 the difference between the product and reactant stoichiometric coefficients, $\Delta\nu$,

$$\Delta\nu = \nu_3 + \nu_4 - \nu_1 - \nu_2$$

appears. When $\Delta\nu$ is nonzero, the right-hand sides of both equations will contain the total pressure raised to some power. Consequently, the units for pressure, both total and partial, enter into chemical equilibrium calculations. *It is a well-established convention to deal with pressures in units of atmospheres for all computations.* It is also evident from Eqs. 10-17 and 10-18 that for reactions with nonzero $\Delta\nu$, the equilibrium chemical composition is a function of both total pressure and absolute temperature.

We can now return to the original question; Given the pressure and temperature of the gas mixture in Fig. 10.2, is the system in chemical equilibrium? The five mole numbers provide the partial pressures for the law of mass action as written in Eq. 10-12, or the total mole number N in Eq. 10-17, or the mole fractions in Eq. 10-18. The right-hand side of any one of these equations may be calculated. (All three give the same value, of course.) Call that value K_p'. From the given temperature of the system, K_p can be calculated or found from tables. Whether or not the system is in chemical equilibrium depends on the agreement between K_p' and K_p. How are we to decide when the agreement between the two is "good enough"? The answer comes by way of van't Hoff's equation. For when we let

$$dK_p = K_p' - K_p$$

then the discrepancy between the two can be related to a temperature difference, dT, by arranging Eq. 10-14 in the form

$$dT = \frac{RT^2}{\Delta H} \frac{K_p' - K_p}{K_p}.$$

If dT is small compared to T, the system is in equilibrium.

10.4 THE LAW OF MASS ACTION FOR OTHER THAN ISOLATED SYSTEMS

Equations 10-12, 10-17, and 10-18, which are equivalent forms of the law, were secured by deriving the conditions that are necessary to produce stationary total entropy in an isolated system in which a reversible reaction of the type shown in Eq. 10-2 can take place. Isolated systems are a special case. What is the criterion for systems that have constraints other than constant volume and constant energy? In Fig. 10.3, suppose the isolated system in (*a*) has come to a state of chemical equilibrium with pressure P and temperature T. Suppose, further that the ambient pressure and temperature are likewise P and T. Now imagine that the insulating blanket is removed, as in (*b*). The system is now constrained to a fixed volume and a fixed temperature, namely that of the surroundings. Suppose, further, that the pin is removed, producing (*c*), which is constrained to the pressure and temperature of the surroundings. But since the blanket and the pin can be removed reversibly, carrying out either or both has no effect on the state of the system. It can be concluded, then, that the *law of mass action, as stated in Eqs. 10-12, 10-17, and 10-18, applies to any and all systems of constraints imposed on a mixture of ideal, reacting gases.*

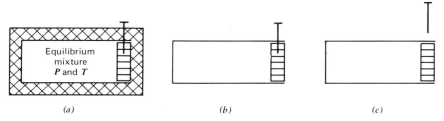

Figure 10.3 Changing the constraints on a system in chemical equilibrium does not change the state of the system. In (a) the system is insulated, and the piston pinned. In (b) the insulation has been removed. In (c) the pin has been removed.

The conclusion drawn from Fig. 10.3, that the criterion for chemical equilibrium is independent of the constraints imposed on the system, should not be confused with the events that will take place in the four systems illustrated in Fig. 10.4. The initial state in each cylinder is a gaseous mixture, $H_2 + O_2$, at P_o and T_o, which are also the ambient pressure and temperature. All four systems are initially in thermodynamic equilibrium, which is to say that slight heating, cooling, compression, or expansion will not produce a large-scale effect. However, when a tiny quantity of energy is added in the form of a spark, a violent explosion occurs, indicating that the initial state is not one of chemical equilibrium.

The constraints on the four systems are different: (a) is isolated; in (b) the piston is free to move, but heat transfer cannot occur; (c) is the reverse of (b); and in (d) both heat transfer and piston motion are possible. Because the constraints are not identical, the final states reached by the four systems will each be different from the others. Each system will reach chemical equilibrium in the final state; the law of mass action applies to each.

10.5 ENTROPY AS A MAXIMUM
The law of mass action is the condition for a stationary entropy. The mathematical demonstration that the entropy is a maximum necessitates proving $d^2 S$ is negative for all possible variations of the system, consistent with the constraints. The system is isolated, as before. Starting with

$$S = \sum_i n_i s_i$$

differentiating twice,

$$d^2 S = \sum_i (s_i \, d^2 n_i + n_i \, d^2 s_i + 2 \, dn_i \, ds_i)$$ \hspace{2cm} 10-19

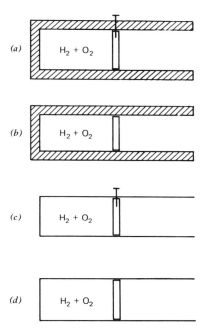

Figure 10.4 Four systems with identical initial states but different constraints. The final states will not be identical.

For each s_i, by the first law,

$$ds_i = \frac{du_i + p_i \, dv_i}{T}$$

and

$$d^2 s_i = - ds_i \frac{dT}{T} + \frac{1}{T} (d^2 u_i + dp_i \, dv_i + p_i \, d^2 v_i) \qquad 10\text{-}20$$

Introducing the constraints, and differentiating twice

$$U = \sum_i n_i u_i = \text{constant}$$

leads to

$$\sum_i n_i \, d^2 u_i = - \sum_i (2 \, dn_i \, du_i + u_i \, d^2 n_i) \qquad 10\text{-}21$$

and

$$V = n_i v_i, \qquad \text{for all } i$$

leads to

$$n_i \, d^2 v_i = -(2 \, dv_i \, dn_i + v_i \, d^2 n_i), \qquad \text{for all } i \qquad \text{10-22}$$

Now substitute the expression for $d^2 s_i$, Eq. 10-20, into Eq. 10-19. Then eliminate the $d^2 u_i$ and $d^2 v_i$ terms using Eqs. 10-21 and 10-22. Making use of the condition for stationary entropy (the law of mass action, Eq. 10-10), the resulting cancellations reduce the entropy derivative to

$$T \, d^2 S = \sum_i n_i (dp_i \, dv_i - ds_i \, dT) \qquad \text{10-23}$$

For ideal gases,

$$ds_i = c_{vi} \frac{dT}{T} + R \frac{dv_i}{v_i}$$

and

$$dp_i = p_i \frac{dT}{T} - p_i \frac{dv_i}{v_i}$$

Substituting for ds_i and dp_i in Eq. 10-23, the result is

$$T \, d^2 S = -\sum_i n_i \left[\frac{c_{vi}}{T} (dT)^2 + \frac{p_i}{v_i} (dv_i)^2 \right] \qquad \text{10-24}$$

The right-hand side is always negative; entropy is a maximum when the law of mass action is satisfied.

(Note: The law of mass action is usually derived as the condition for stationary free energy (Gibbs function) in a system that remains at constant pressure and temperature. Use of the free energy function leads to a more elegant mathematical treatment, and the fact that the free energy is a minimum provides useful information. The free energy is, however, a derived function, following after the formulation of the entropy function. The author prefers the use of entropy, as above, since it makes a more direct appeal to the second law. The free energy approach to the study of chemical equilibrium is presented in Appendix D.)

10.6 CHEMICAL EQUILIBRIUM IN MULTIREACTION SYSTEMS

Systems in which chemical equilibrium is reached under the action of a single reaction are the exception. Multireaction systems are the rule. Combustion of a reactant mixture containing hydrogen and oxygen leads to a product mixture in which four reactions can occur:

$$H_2O \rightleftharpoons H_2 + \tfrac{1}{2}O_2$$
$$O_2 \rightleftharpoons 2O$$
$$H_2 \rightleftharpoons 2H$$
$$H + O \rightleftharpoons OH$$

When the reactant mixture contains a hydrocarbon and oxygen, a fifth reaction can be added to the above four:

$$CO_2 \rightleftharpoons CO + \tfrac{1}{2}O_2$$

A sixth and a seventh reaction can arise when the reactants comprise a hydrocarbon and air:

$$N_2 \rightleftharpoons 2N$$
$$N_2 + O_2 \rightleftharpoons 2NO$$

It is a straightforward exercise to show that a law of mass action is associated with each and every reaction in multireaction systems. The proof is left to the reader. See Problems 10.1 and 10.2.

At sufficiently high temperatures, ionization reactions will be present, for example,

$$O \rightleftharpoons O^+ + e^-$$

Combustion seldom produces sufficient quantities of ionized gases to influence the adiabatic flame temperature appreciably.

10.7 VALUES OF THE REACTION CONSTANTS

Table E.5, is a tabulation of $\log_{10} K_p$ against T for eight formation reactions. Since formation reactions may be combined to produce new reactions, K_p values may be manipulated correspondingly. For example, to find K_p for the following reaction

$$H_2 + \tfrac{1}{2}O_2 \rightleftharpoons H_2O$$

at 298 K, the definition of K_p is

$$K_p = \frac{p_{H_2O}}{p_{H_2}\sqrt{p_{O_2}}}$$

Rewrite that reaction in terms of formation reactions

$$K_p = \left(\frac{p_{H_2O}}{p_H^2\sqrt{p_O}}\right)_I \left(\frac{p_H^2}{p_{H_2}}\right)_{II} \left(\frac{p_O}{p_{O_2}}\right)_{III}^{1/2}$$

and

$$\log_{10}K_p = \log_{10}K_{pI} + \log_{10}K_{pII} + \tfrac{1}{2}\log_{10}K_{pIII}$$

Values for the three quantities on the right are found in Table E.5, and

$$\log_{10}K_p = 151.5 - 71.2 - \frac{80.6}{2} = 40.0$$

10.8 LECHATELIER'S EQUATIONS

When the conditions, such as P and T, under which a system is in chemical equilibrium change, the composition of the system will, in general, change also, provided equilibrium prevails. The direction and magnitude of the alterations in composition are simple to evalute in single reaction systems. Multireaction systems can be analyzed, but the results involve cumbersome algebraic expressions that will not be developed here.

Let the single reaction be represented by

$$\nu_1 A_1 + \nu_2 A_2 \rightleftharpoons \nu_3 A_3 + \nu_4 A_4 \qquad\qquad 10\text{-}25$$

for which

$$\Delta\nu = \nu_3 + \nu_4 - \nu_1 - \nu_2$$

and

$$\Delta H = \nu_3 h_3 + \nu_4 h_4 - \nu_1 h_1 - \nu_2 h_2$$

The reactive gases are present in amounts n_1, n_2, n_3, and n_4. In addition, n_x moles of an inert gas are present.

Now, if P changes to $P + dP$, or T changes to $T + dT$, or n_x changes to $n_x + dn_x$, how will n_1, n_2, n_3, and n_4 change? Since the changes in the mole numbers of the four reactive gases are not independent but are related to each other by Eq. 10-8 (conservation of matter), the composition change can be understood by deriving expressions for the change in one of the four, say n_1. What we want, therefore, are equivalent forms for the differentials

$$\left(\frac{\partial n_1}{\partial P}\right)_{T,n_x}, \qquad \left(\frac{\partial n_1}{\partial T}\right)_{P,n_x}, \qquad \left(\frac{\partial n_1}{\partial n_x}\right)_{P,T}$$

Starting with the law of mass action in the form given by Eq. 10-17 differentiate and obtain

$$\frac{dK_p}{K_p} = \nu_3 \frac{dn_3}{n_3} + \nu_4 \frac{dn_4}{n_4} - \nu_1 \frac{dn_1}{n_1} - \nu_2 \frac{dn_2}{n_2} + \Delta\nu \frac{dP}{P} - \Delta\nu \frac{dN}{N}$$

Eliminating K_p with van't Hoff's equation, Eq. 10-14, and eliminating n_2, n_3, and n_4 with Eq. 10-8, the result is

$$\frac{1}{\Psi} \frac{dn_1}{\nu_1} = -\frac{\Delta H}{RT^2} dT + \Delta\nu \frac{dP}{P} - \Delta\nu \frac{dn_x}{N} \qquad\qquad 10\text{-}26$$

in which the quantity Ψ has the form

$$\frac{1}{\Psi} = \frac{\nu_1^2}{x_1} + \frac{\nu_2^2}{x_2} + \frac{\nu_3^2}{x_3} + \frac{\nu_4^2}{x_4} - (\Delta\nu)^2 \qquad\qquad 10\text{-}27$$

In Appendix D it is shown that Ψ must be positive for all reactions. Consequently, the directions in which n_1 changes with P, T, and n_x can be deduced from Eq. 10-26

$$\left(\frac{\partial n_1}{\partial P}\right)_{T,n_x} \sim + \Delta\nu \qquad\qquad 10\text{-}28$$

$$\left(\frac{\partial n_1}{\partial T}\right)_{P,n_x} \sim - \Delta H \qquad\qquad 10\text{-}29$$

$$\left(\frac{\partial n_1}{\partial n_x}\right)_{P,T} \sim - \Delta\nu \qquad\qquad 10\text{-}30$$

These are LeChatelier's equations. They demonstrate the stability of gaseous systems in chemical equilibrium. The key factors are the signs of $\Delta\nu$ and ΔH.

The response of a stable system is always to offset changes imposed by the surroundings. For example, suppose $\Delta \nu$ is positive for reaction 10-25. Then the total mole number, N, will increase (decrease) as 10-25 proceeds in the forward (reverse) direction. According to Eq. 10-28, n_1 will increase when $\Delta \nu$ and dP are both positive. To increase n_1, reaction 10-25 must proceed in the reverse direction. This will decrease N. Equation 10-28 states, then, that P and N move in opposite directions. The system is stable. The system is unstable when P and N both increase (or decrease), for then each increases from the action of the other, and P will continue to increase without further interaction with the surroundings.

If ΔH is positive, the system absorbs energy when reaction 10-25 proceeds in the forward direction. Then if ΔH is positive, and T increases, the system must absorb energy, and the reaction must proceed to the right, which means n_1 must decrease, as predicted by Eq. 10-29. If n_1 were to increase with an increase in T, the reaction would release energy, and the system would steadily become warmer, without further addition of energy from the surroundings. That sort of behavior is unstable.

The mole change, $\Delta \nu$, may be zero, as in the so-called 'water gas' reaction

$$CO + H_2O \rightleftharpoons CO_2 + H_2$$

For all other reactions encountered in the product mixtures resulting from combustion, $\Delta \nu$ and ΔH have the same sign. In a single reaction system, pressure and temperature have opposite effects on the equilibrium composition.

With regard to the magnitude of the change in equilibrium composition when P, T, or n_x are altered, the quantity Ψ enters. A certain amount of symmetry can be achieved by introducing a new variable, the *"extent of reaction,"* ϵ, so the mole numbers for the reactive gases can be written

$$n_1 = \nu_1(1 - \epsilon) + a_1, \qquad n_3 = \nu_3 \epsilon + a_3$$
$$n_2 = \nu_2(1 - \epsilon) + a_2, \qquad n_4 = \nu_4 \epsilon + a_4 \qquad \text{10-31}$$

Clearly, ϵ is restricted to values between zero and unity. The a_i are integration constants. A moment's reflection leads to the conclusion that a_1 or a_2 must be zero (or both may be zero), and similarly for a_3 and a_4. The a_i denote the "excess" amount of any gas present. If there is excess gas present, in the sense that there is not a sufficient amount of its partner in the system to permit reaction, the excess gas merely absorbs heat as the system temperature is increased, for example. If there is no excess of any gas, more of the energy added from the surroundings is available to cause chemical reaction; the system has greater chemical "mobility." Setting the $a_i = 0$ in Eq. 10-31, from Eq. 10-27,

$$\Psi = \frac{(1 - \epsilon)\epsilon}{(\nu_1 + \nu_2)(\nu_3 + \nu_4)}$$

and Ψ will be a maximum when $\epsilon = \frac{1}{2}$, in the middle of the allowable range, regardless of pressure and/or temperature,

$$\Psi_{max} = \frac{1}{4(\nu_1 + \nu_2)(\nu_3 + \nu_4)}$$

10.9 THREE ELEMENTARY EXAMPLES

Three examples of chemical equilibrium in elementary systems are discussed next to illustrate how equilibrium compositions can be computed and to show the magnitude of the energy effects that can result when chemical equilibrium prevails.

Example 1. How much energy is required to heat 1 mol of hydrogen from 298 to 4000 K at constant pressure?

(*a*) The elementary solution ignores dissociation. The required amount of energy is the enthalpy difference between the final and initial states:

$$Q = h_{H_2}(4000) - h_{H_2}(298) = 417 - 291 = 126 \text{ kJ}$$

(*b*) The refined solution takes into account possible dissociation, and chemical equilibrium through

$$2H \rightleftharpoons H_2$$

At 298 K, referring to Table E.5, Appendix E,

$$K_p = \frac{p_{H_2}}{p_H^2} = 10^{71.2}$$

If the total pressure is, for example, 1 atm, then

$$p_H + p_{H_2} = 1$$

and we conclude that $p_{H_2} = 1$ atm, and $p_H = 10^{-35}$ atm; at 298 K hydrogen exists entirely in the molecular state.

To derive an expression for the mole number of atomic hydrogen, n_H, write conservation of matter

$$n_H + 2n_{H_2} = 2$$

and the law of mass action, Eq. 10-17,

$$K_p = \frac{n_{H_2O}(n_H + n_{H_2})}{n_H^2 P} \qquad (P \text{ in atm})$$

or

$$n_H = \frac{2}{\sqrt{1 + 4PK_p}}$$

which demonstrates that n_H is a function of pressure and temperature. As P approaches zero n_H approaches 2; as P becomes very large, n_H approaches zero. At 4000 K, $\log_{10} K_p = -.4012$, and the mole numbers, as functions of total pressure are

P(atm)	n_H	n_{H_2}
0.01	1.98	0.01
0.1	1.86	0.07
1.0	1.24	0.38
10.0	0.487	0.76
100.0	0.158	0.92

The required energy is the enthalpy difference between the final and initial states,

$$Q = (n_H h_H + n_{H_2} h_{H_2})_{4000} - h_{H_2}(298)$$

Since the enthalpy tables in Appendix E are constructed with datum values set to include heats of formation, Q may be evaluated directly. The variation of Q with total pressure P is

P(atm)	Q(kJ)
0.01	582
0.1	556
1.0	414
10.0	238
100.0	163
∞	126

The influence of the total pressure is enormous.

An alternative for the energy calculation could proceed as follows: imagine that the hydrogen is heated to 4000 K, without dissociation. This requires, as we have seen, 126 kJ. Suppose the total pressure is 1 atm. The equilibrium mixture contains 0.38 mol of H_2, so that 0.62 mol must be dissociated. At 4000 K the energy required to dissociate 1 mol of H_2 at constant pressure is

$$\Delta H = (2h_H - h_{H_2})\ 4000\ K$$
$$= 2 \times 440 - 417$$
$$= 463\ kJ$$

To dissociate 0.62 mol will then require

$$0.62 \times 463 = 287 \text{ kJ}$$

Consequently, the total energy requirement for heating at 1 atm total pressure is $126 + 287 = 413$ kJ.

Example 2. How much energy is required to heat 1 mol of hydrogen from 298 to 4000 K at constant volume?

The solution with dissociation included combines conservation of matter

$$n_H + 2n_{H_2} = 2$$

with the law of mass action

$$K_p = \frac{n_{H_2} N}{n_H^2 P}$$

where P and N are the total pressure and total mole number in the final state. Since the volume is constant,

$$\frac{N}{P} = \frac{N_i T_i}{P_i T}$$

where subscript i denotes the inital state, and T is the final state. Eliminating n_{H_2}

$$n_H = \frac{A}{4} \left[\sqrt{1 + \frac{16}{A}} - 1 \right], \qquad \text{where } A = \frac{T_i}{P_i T K_p}$$

The results, for various values of P_i are

P_i(atm)	n_H	n_{H_2}	Q(kJ)
0.01	1.68	0.16	456
0.1	0.98	0.51	310
1.0	0.39	0.81	192
10.0	0.13	0.93	126
100.0	0.04	0.98	105

$$Q = (n_H u_H + n_{H_2} u_{H_2})_{4000} - u_{H_2} (298)$$

Example 3. What is the equilibrium composition of a mixture of 1 mole of hydrogen and n_x moles of an inert gas, at 4000 K and 1 atm?

Combining conservation of matter

$$n_H + 2n_{H_2} = 2$$

and the law of mass action

$$K_p = \frac{n_{H_2}}{n_H^2} \frac{n_H + n_{H_2} + n_x}{P}$$

and solving for n_H,

$$n_H = \frac{-n_x + \sqrt{n_x^2 + 4(1 + n_x)(1 + 4\,PK_p)}}{1 + 4\,PK_p}$$

The variations of n_H and n_{H_2} with n_x are

n_x	n_H	n_{H_2}
0	1.24	0.38
1	1.41	0.30
2	1.51	0.25
3	1.58	0.21

To confirm these results with the prediction of Eq. 10-30, if $n_1 = n_H$ in that equation, then reaction 10-25 must be written

$$2H \rightleftharpoons H_2$$

for which $\Delta\nu = -1$, and Eq. 10-30 reads

$$\left(\frac{\partial n_H}{\partial n_x}\right)_{P,T} > 0$$

What is the physical explanation for the increase in n_H that results from increasing n_x? Imagine the system in equilibrium with no inert gas present. During an interval of time, a number of molecules, m, dissociate, and an equal number are formed by successful atom–atom collisions. The number m depends on the number of molecules and atoms present and on the fraction of molecules that are sufficiently energetic and the fraction of collisions that are successful. Both fractions are fixed by the prevailing temperature. (That is the main conclusion to be drawn from the law of mass action.)

Now add inert gas. Since P and T remain constant, the volume of the system must increase. Suppose, for simplicity, that the volume is doubled, instantaneously. Now examine what happens during successive time intervals. During the first instant, m molecules will dissociate, as before, but only $m/2$ molecules will be formed by collisions; the average distance between atoms has doubled, which halves the collision rate. During the second time interval, fewer molecules will dissociate (there are fewer of them), but more will be produced, as compared to the production during the first time interval, since there are more atoms present. (The one-half factor no longer applies, since the volume will continually change in response to the number of particles in the system.) As time passes the system will move to a new equilibrium composition that will contain fewer molecules and more atoms than were present before the inert gas was introduced.

10.10 SUMMARY

Chemical equilibrium is established when the rates of dissociation and formation of each reactive gas in the system balance each other. The law of mass action, Eq. 10-12 or 10-17 or 10-18, applies to each reversible reaction. The equation of van't Hoff, Eq. 10-14, is useful for determining whether the discrepancy between the reaction constant K_p and its equivalent, computed from the given chemical composition, is sufficiently small to be considered negligible.

The reaction constants are functions of absolute temperature. Since the heats of reaction for reversible reactions are weakly dependent on temperature, van't Hoff's equation in the form

$$\frac{d(\ln K_p)}{d\left(\frac{1}{T}\right)} = -\frac{\Delta H}{R} \qquad \text{10-32}$$

can be integrated, taking ΔH constant, to give an approximate relationship between K_p and T.

The numerical value of K_p depends on the choice of units for pressure and the choice of stoichiometric coefficients. *Pressures must be in units of atmospheres.* The stoichiometric coefficients for formation reactions are fixed by convention. For other reversible reactions, it is best to use the reactions as written in Table C.2. Appendix C. Reversing the "reactants" and "products" in a reversible reaction inverts the reaction constant and changes the algebraic signs of $\Delta \nu$ and ΔH.

The influences of pressure, temperature, and inert gas content on the equilibrium composition can be determined from LeChatelier's equations, Eqs. 10-28, 10-29, and 10-30, for single reaction systems. Multireaction systems exhibit more complicated dependencies. For example, when CO_2 is heated, dissociation produces CO and O_2, the latter dissociating to produce O. At low temperatures, only CO_2 will be present, while at high temperatures, only CO and O will be present. Equilibrium is maintained by

$$CO_2 \rightleftharpoons CO + \tfrac{1}{2}O_2$$
$$2O \rightleftharpoons O_2$$

The amount of O_2 must therefore reach a maximum value at some intermediate temperature, increasing with T below that temperature and decreasing above it.

PROBLEMS

10.1 For a system in which equilibrium is maintained by two independent reactions

$$\nu_1 A_1 + \nu_2 A_2 \rightleftharpoons \nu_3 A_3$$
$$\nu_4 A_4 + \nu_5 A_5 \rightleftharpoons \nu_6 A_6$$

show that the entropy of an isolated system will be stationary when the law of mass action for both equations is satisfied.

10.2 For a system that incorporates two coupled reactions

$$\nu_1 A_1 + \nu_2 A_2 \rightleftharpoons \nu_3 A_3$$
$$\nu_1 A_1 \rightleftharpoons \nu_4 A_4$$

show that the entropy of an isolated system will be stationary when the law of mass action is satisfied for both reactions.

10.3 A mixture of 1 mol each of O_2 and N_2, initially at 300 K and 0.5 atm, is heated at constant pressure by the addition of 750 kJ. What is the final temperature if the final state is an equilibrium mixture of O_2, N_2, and O?

10.4 One mole each of O_2 and N_2 are placed in equal-sized compartments in a rigid vessel, separated by a rigid partition. The initial pressure and temperature in both compartments is 0.1 atm and 300 K. The gases are heated to 4000 K. How much energy is required if the oxygen dissociates but the nitrogen does not?
What will be the final state of the system if the partition now ruptures and heat transfer to the surroundings is prevented by an insulating jacket?

11

CHEMICAL EQUILIBRIUM IN C/H/O/N SYSTEMS

11.1 INTRODUCTION

The subject of this chapter is the computation of chemical equilibrium mixture compositions for systems containing C, H, O, and N atoms. Mixtures of this type are encountered in the working fluid of internal combustion engines and in rockets. Our discussion will be restricted to the following relatively simple mixtures: six species, with P and T known; six species, with V and T known; ten species; and further extension, to 12 or more species.

There are two reasons for making equilibrium mixture calculations. First, the pressure and temperature that follow combustion in internal combustion engines, whether two or four stroke, Diesel or Otto, depend on the character of the mixture of product gases. (The situation is complicated by the fact that the character of the mixture depends on prevailing pressure and temperature. These matters are discussed in Chapter 13.) The temperature reached in the combustion chamber of a rocket depends on the composition of the products of combustion. On the other hand, the work produced by the working fluid in an engine and the thrust developed in a rocket depend on pressures and temperatures. Hence, more accurate work and thrust estimates are obtained when the assumption of complete combustion is replaced by the assumption of prevailing chemical equilibrium. At the same time, there is a substantial increase in the amount of computation required; equilibrium mixture calculations can become complicated and involved. In short, the mathematical model of combustion, whatever the application or purpose, improves when we allow for equilibrium producing reactions in the product mixture.

The other reason for making such calculations is for the purpose of examining the chemical nature of the product mixture. The character of exhaust emissions from internal combustion engines has attracted considerable attention in recent years owing to the undesirable properties of carbon monoxide, CO, and nitric oxide, NO. The latter is a stable molecule that accumulates in the atmosphere and very likely remains there. The amounts of CO and NO produced by an engine depends on how it operates and what fuel is burned. In order to estimate these amounts, their presence in the product gas mixture must be assumed and allowed for in equilibrium calculations. That calculation, however, may be only a crude estimate; in real situations, equilibrium may not be reached because of the kinetics, that is, the time rates of reactions. Under these conditions, which are by no means uncommon, calculation of the product mixture composition is considerably increased in complexity, and the results difficult to interpret, owing to our limited knowledge about the precise behavior of reacting gases in circumstances that are inaccessible for measuring instruments.

Two computational schemes are discussed in this chapter, one with six species and the other with ten. These are simple mixtures when it comes to calculating equilibrium composition. They are considered for two reasons. Because they are simple, they provide a good introduction to a calculational problem that can become extremely challenging when the numbers of species are increased. The second reason for considering these simple systems is that in some instances the information they provide about the chemical and thermodynamic nature of the products of combustion is "good enough" for engineering purposes, especially when it is recognized that the conventional models for internal combustion engines and rockets are highly idealized representations of complicated machines and processes that are presently only crudely understood.

11.2 SIX SPECIES, WITH P AND T KNOWN

The problem to be discussed can be stated as follows: If known quantities of carbon, hydrogen, oxygen, and nitrogen form a mixture of gases at known values of pressure and temperature, what will be the equilibrium composition if the species in the mixture are limited to

$$N(1)CO_2 + N(2)CO + N(3)O_2 + N(4)H_2O + N(5)H_2 + N(6)N_2$$

that is, what are the values of the six mole numbers if the mixture is in chemical equilibrium?

The selection of these six species of gases is arbitrary. Other selections could be made, and we can consider mixtures with more than six species. This degree of arbitrariness and selection is always present in problems involving chemical equilibrium; a decision must be made as to what kinds of molecules and atoms will be included in the solution, and that decision must be made at the outset of the solution. The choice made above is the logical first step beyond the assumption of complete combustion and would be appropriate for any internal combustion engine.

For six unknowns, we require six equations. Let AC, AH, AO, and AN denote the *abundancies* of carbon, hydrogen, oxygen, and nitrogen *atoms* in the system, expressed in units of moles or kilomoles. Then four of the six equations will be statements of conservation of the four types of atoms. The remaining equations are law of mass action equations for the reactions that maintain the mixture in a state of chemical equilibrium. For the six gases in the selected mixture, there are three possible reactions:

$$CO + \tfrac{1}{2}O_2 \rightleftharpoons CO_2 \qquad \text{(reaction constant } K2)$$
$$H_2 + \tfrac{1}{2}O_2 \rightleftharpoons H_2O \qquad \text{(reaction constant } K1)$$
$$CO + H_2O \rightleftharpoons CO_2 + H_2$$

However, this is not an independent set of reactions; each is some linear combination of the other two. The remaining two equations, which complete the set of six, may be mass action equations for any two of the three reactions. We shall choose the first two, denoting the reaction constants by $K1$ and $K2$, as indicated.

The six equations can now be set down:

$$
\left.
\begin{aligned}
AC &= N(1) + N(2)\\
AH &= 2[N(4) + N(5)]\\
AO &= N(2) + N(4) + 2[N(1) + N(3)]\\
AN &= 2N(6)\\[6pt]
K2 &= \frac{N(1)}{N(2)\sqrt{N(3)}}\sqrt{\frac{NT}{P}}\\[6pt]
K1 &= \frac{N(4)}{N(5)\sqrt{N(3)}}\sqrt{\frac{NT}{P}}
\end{aligned}
\right\}
\qquad 11\text{-}1
$$

The first four are statements for conservation of matter. The last two are mass action equations in the form of Eq. 10-17. Pressure, in units of atmospheres, is represented by P, NT is the total mole number, and temperature T fixes the values of $K1$ and $K2$, through expressions of the type given in Eq. 10-16, for which the a, b, c, and d coefficients may be found in Table C.2. In Eqs. 11-1, then, the left-hand sides are known, as is P.

That set of equations can be solved in a number of ways. We shall follow out one solution based on the half-interval search strategy. Actually, there is a double search, for we shall treat the total mole number NT as a pseudo-unknown. We can, of course, eliminate NT in terms of the six mole numbers, but that will not be done in this solution.

To start with, we know $N(6) = AN/2$, and we continue as follows:

1. Choose a "reasonable" guess value for NT. If, for example, the mixture of gases is the result of combustion, the complete combustion value for NT could be used as the first guess. The choice is not critical; the solution will

converge for wildly absurd initial values. In the course of the solution, we shall secure computed values for NT, and the calculations will cease when the assumed and computed values agree within some tolerable error.

2. Start the half-interval search with $N(1)$ in the middle of its allowable range,

$$N(1) = \frac{AC}{2}$$

3. In the following order, compute

$$\left. \begin{array}{l} N(2) = AC - N(1) \\[2mm] N(3) = \left[\dfrac{N(1)}{N(2)\,K2} \right]^2 \dfrac{NT}{P} \\[3mm] N(4) = AO - 2[N(1) + N(3)] - N(2) \\[2mm] N(5) = \dfrac{AH}{2} - N(4) \end{array} \right\} \qquad 11\text{-}2$$

4. With the values of $N(3)$, $N(4)$, and $N(5)$ just calculated, compute $K1'$ (the last equation in 11-1 was not used in 11-2)

$$K1' = \frac{N(4)}{N(5)\sqrt{N(3)}} \sqrt{\frac{NT}{P}} \qquad\qquad 11\text{-}3$$

5. Compare $K1$ and $K1'$. The algebraic sign of their difference informs us which sign should be selected in computing the next value for $N(1)$

$$N(1) = N(1) \pm \frac{AC}{2^{J+1}} \qquad\qquad 11\text{-}4$$

where J is an integer variable, set equal to unity at the start and increased by unity for each iteration.

6. The iterations on $N(1)$ are continued until $K1'$ and $K1$ agree within some acceptable error.

7. Compute $NT' = N(1) + \ldots + N(6)$, and compare with NT. If the agreement is within acceptable error, the computation is complete. Otherwise, set $NT = NTP$, and return to step 2.

The logic that guides the half-interval search is contained in the functional relationship between $N(1)$ and $K1'$, which can be ascertained by differentiating Eqs. 11-2 and 11-3, treating P, $K2$, and NT as constants, and writing each differential in terms of $dN(1)$; thus,

$$dN(2) = -dN(1)$$

$$dN(3) = 2N(3)\left[\frac{1}{N(1)} + \frac{1}{N(2)}\right]dN(1)$$

$$dN(4) = -\left[1 + 4N(3)\left(\frac{1}{N(1)} + \frac{1}{N(2)}\right)\right]dN(1) \qquad\qquad 11\text{-}5$$

$$dN(5) = 1 + 4N(3)\left(\frac{1}{N(1)} + \frac{1}{N(2)}\right)dN(1)$$

$$\frac{d(K1')}{K1'} = -\left\{\left[\frac{1}{N(4)} + \frac{1}{N(5)}\right]\left[1 + 4N(3)\left(\frac{1}{N(1)} + \frac{1}{N(2)}\right)\right] + \frac{1}{N(1)} + \frac{1}{N(2)}\right\}dN(1)$$

Since the mole numbers are positive, it follows that $N(1)$ and $K1'$ change oppositely. But a glance at Eqs. 11-2 shows that $N(4)$ and $N(5)$ can turn up negative, and since negative mole numbers are physically impossible, they must be avoided like the plague. Equations 11-5 indicate what is to be done to $N(1)$ in that eventuality, and the search logic can be written as follows:

If $N(4)$ is negative, decrease $N(1)$;
If $N(5)$ is negative, increase $N(1)$;
If $K1'$ is greater than $K1$, increase $N(1)$;
If $K1'$ is less than $K1$, decrease $N(1)$.

The solution outlined here involves two search routines, one on $N(1)$ and the other on NT. We need to establish criteria for halting both searches when the errors are no longer significant. There are two ways to treat the half-interval search on $N(1)$. The difference between $K1'$ and $K1$ can be converted to a temperature error, DT, by way of van't Hoff's equation, Eq. 10-14; that is,

$$DT = \left|\frac{K1' - K1}{K1\Delta H1}\right|RT^2 \qquad\qquad 11\text{-}6$$

where $\Delta H1$ is the heat of reaction

$$\Delta H1 = (h_{H_2O} - h_{H_2} - \tfrac{1}{2}h_{O_2})_T$$

Since heats of reaction are weak functions of T, the value at 300 K can be used,

$$\Delta H1 = -214\ 800\ \text{kJ}$$

Thus the iterations on $N(1)$ can be continued until DT becomes less than some preassigned error, say $100°C$, or $5°C$, or whatever, depending on the desired accuracy.

The other way to treat the iteration on $N(1)$ is to examine the differences in mole fraction for each species (except N_2, which is fixed) between successive iterations. Thus, if $Z(I)$ denotes the mole fractions associated with the previous iteration, the half-interval search on $N(1)$ may be halted when

$$\frac{|N(I) - Z(I)|}{NT} \leqslant \text{error}, \qquad \text{for } I = 1,5 \qquad\qquad 11\text{-}7$$

where some preassigned limit, say 0.01 or 0.001, is set for the quantity *error*, depending on the desired or required accuracy. The choice between these two methods is made depending on whether a temperature error or mole fraction errors seem to be most appropriate for the ultimate purposes of the computation.

The iteration on the total mole number NT can be halted when

$$\frac{|NT - NT'|}{NT} \leqslant \text{error} \qquad\qquad 11\text{-}8$$

and here the *error* might be 0.02 or 0.01 or the like.

Because of the manner in which NT and P occur in Eqs. 11-1, it is worth noting that when a set of six mole numbers $N(1), \ldots, N(6)$ satisfy Eq. 11-7, those mole numbers are an equilibrium mixture for the given T and for a pressure P', which is given by

$$P' = \frac{NT'}{NT} P$$

It might be supposed that the solution time can be advantageously shortened by updating the value of NT at the conclusion of each separate iteration on $N(1)$. This may or may not "work"; when it does not work the program becomes mired down in what is called "whipping," a back-and-forth oscillation in NT which, under some circumstances, will not die out. The program described here involves a pair of nested iterations, and as a general rule, they should be treated in strict order, from inner to outer. In the present case, the logic of the program was developed by differentiating Eqs. 11-2, treating NT as a constant. If, instead, NT is changed each time an iteration on $N(1)$ is completed, that logic may or may not be valid.

By way of further explanation, a flow chart of the program is outlined in Fig. 11.1. The iterations on $N(1)$ are stopped when the error in Eq. 11-7 is less than 0.001, and the iterations on NT are stopped when the error in Eq. 11-8 is less than 0.02. These

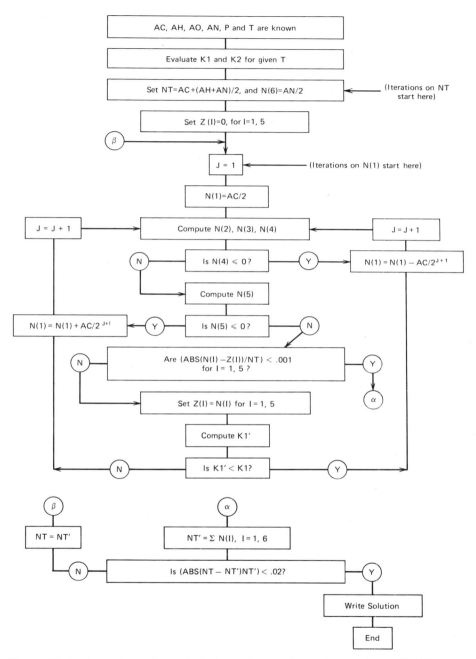

Figure 11.1 Flow chart for calculation of the six mole numbers, $N(1)CO_2 + N(2)CO + N(3)O_2 + N(4)H_2O + N(5)H_2 + N(6)N_2$. Given: pressure P, temperature T, and the abundancies AC, AH, AO, and AN.

error limits are mentioned only to explain those numbers where they occur in the flow chart. Other limits may be more appropriate, according to circumstances.

Three final comments on the program and the flow chart follow:

1. The program will operate satisfactorily for $AN = 0$, that is, for a system containing only C, H, and O atoms.
2. The program will not work for $AO = 0$ or $AC = 0$ or $AH = 0$.
3. The program will not work for AO less than AC. (Why?)

11.3 SIX SPECIES, WITH V AND T KNOWN

As in Section 11.2, the problem is to compute the mole numbers in the mixture

$$N(1)CO_2 + N(2)CO + N(3)O_2 + N(4)H_2O + N(5)H_2 + N(6)N_2$$

given the four abundancies, the volume V, and the temperature T. The six equations listed in Eqs. 11-1 provide the solution. However, the quantity P/NT which appears in the mass action statements is now known,

$$\frac{P}{NT} = \frac{RT}{V}$$

In the case of constant volume combustion, V would be known from

$$V = \frac{N_r R T_r}{P_r}$$

the subscript r denoting the reactant mixture. Note that the abundancies and N_r must be consistent; that is, the abundancies are to be derived from the composition of the reactant mixture. Note also that P/NT must be dealt with in units of atmospheres per mole or atmosphere per kilomole, depending on the choice of units for the abundancies.

If we now introduce

$$W1 = K1\sqrt{\frac{P}{NT}} \quad \text{and} \quad W2 = K2\sqrt{\frac{P}{NT}}$$

then in place of Eqs. 11-1, the solution for the mole numbers with V and T known is obtained with the following set of equations:

$$N(6) = AN/2$$

$$N(2) = AC - 2N(1)$$

$$N(3) = \left[\frac{N(1)}{N(2)\,W2} \right]^2$$

$$N(4) = AO - N(2) - 2N(1) + N(3)$$

$$N(5) = AH - 2N(4)$$

$$W1' = \frac{N(4)}{N(5)\,\sqrt{N(3)}}$$

$\qquad\qquad$ 11-9

The set is solved by placing $N(1)$ in the middle of its allowable range, after which values for $N(2)$, $N(3)$, $N(4)$, and $N(5)$ are obtained, in that order. Then $W1'$ is compared with the known value of $W1$, and the algebraic sign of the difference directs the next choice for $N(1)$. The logic for this search is identical with the case for P and T known, since

$$\frac{d(W1')}{W1'} = \frac{d(K1')}{K1'}$$

and therefore

If $N(4)$ is negative, decrease $N(1)$;
If $N(5)$ is negative, increase $N(1)$;
If $W1'$ is greater than $W1$, increase $N(1)$;
If $W1'$ is less than $W1$, decrease $N(1)$.

The program will operate satisfactorily with $AN = O$, but not for $AO = O$, or $AC = O$ or $AH = O$. In addition, AO must be larger than AC.

11.4 TEN SPECIES

The mixture of gases is assumed to contain

$$N(1)CO_2 + N(2)CO + N(3)O_2 + N(4)O + N(5)NO +$$
$$N(6)N_2 + N(7)H + N(8)H_2 + N(9)OH + N(10)H_2O$$

and we wish to calculate the 10 mole numbers in the equilibrium state, with either V and T known or P and T known. Ten equations are required. Four are supplied by conservation of the four elements, and the remaining six are mass action statements, derived for the following set of independent reactions:

$$CO + \tfrac{1}{2}O_2 \rightleftharpoons CO_2, \qquad K1$$

$$2O \rightleftharpoons O_2, \qquad K2$$

$$O_2 + N_2 \rightleftharpoons 2NO, \qquad K3$$

$$2H \rightleftharpoons H_2, \qquad K4$$

$$O + H \rightleftharpoons OH, \qquad K5$$

$$H_2 + \tfrac{1}{2}O_2 \rightleftharpoons H_2O, \qquad K6$$

The six reaction constants $K1, \ldots, K6$ are identified with the appropriate reaction, as indicated. When mass action laws are written for these six reactions, in all but the third (in which there is no mole change between reactants and products) the quantity P/NT will occur. As in Section 11.3, we can now introduce W quantities that include the reaction constant and the factor P/NT, so that the solution for the mole numbers can be discussed in a simple fashion, whether V and T or P and T are given. Let

$$\left.\begin{aligned}
W1 &= K1\sqrt{\frac{P}{NT}}\\[6pt]
W2 &= K2\,\frac{P}{NT}\\[6pt]
W4 &= K4\,\frac{P}{NT}\\[6pt]
W5 &= K5\,\frac{P}{NT}\\[6pt]
W6 &= K6\sqrt{\frac{P}{NT}}
\end{aligned}\right\} \qquad\qquad 11\text{-}10$$

If P and T are given, then a value for NT must be assumed and subsequently compared with a computed value, as in Section 11.2 and as indicated in the flow chart, Fig. 11.1.

The 10 equations that form the solution, listed in the order in which they enter into the half-interval search routine, can now be set down. The general notion is to pass from one equation to the next, accumulating values for the 10 mole numbers, which are then checked in a remaining unused equation to see if it is satisfied. The search is continued until the last equation is satisfied. The set of equations is

(a) $AC = N(1) + N(2)$

(b) $W1 = \dfrac{N(1)}{N(2)\sqrt{N(3)}}$

(c) $W2 = \dfrac{N(3)}{N(4)^2}$

(d) $AN = N(5) + 2N(6)$

(e) $K3 = \dfrac{N(5)^2}{N(3)\,N(6)}$

(f) $AO = 2[N(1) + N(3)] + N(2) + N(4) + N(5) + N(9) + N(10)$

(g) $AH = 2[N(8) + N(10)] + N(7) + N(9)$

(h) $W4 = \dfrac{N(8)}{N(7)^2}$

(i) $W5 = \dfrac{N(9)}{N(4)N(7)}$

(j) $W6 = \dfrac{N(10)}{N(8)\sqrt{N(3)}}$

$$\left.\vphantom{\begin{array}{c} a \\ a \\ a \\ a \\ a \\ a \\ a \\ a \\ a \\ a \\ a \end{array}}\right\} \text{11-11}$$

The iterations are started with $N(1) = AC/2$, in the middle of the allowable range, and from the first five equations, we secure values for $N(2)$, $N(3)$, $N(4)$, $N(5)$, and $N(6)$. Now eliminate $N(10)$ between Eqs. 11-11f and 11-11g. The result is

$$-\tfrac{1}{2}N(7) - N(8) + \tfrac{1}{2}N(9) = C \qquad\qquad \text{11-12a}$$

where the term C occupies a position a central importance in the solution. For C we have

$$C = AO - \frac{AH}{2} - 2N(1) - N(2) - 2N(3) - N(4) - N(5) \qquad\qquad \text{11-12b}$$

Now combine Eqs. 11-11h, 11-11i, and 11-12a, solving for $N(7)$, with the result

$$W4N(7)^2 + BN(7) + C = 0 \qquad\qquad \text{11-12c}$$

where

$$B = \tfrac{1}{2}[1 - W5N(4)]$$

and C in Eq. 11-12c is defined in Eq. 11-12b. To avoid ambiguity, the roots of the quadratic equation for $N(7)$ must have opposite signs, and therefore C must be negative. The root for $N(7)$ is then

$$N(7) = \frac{-B + \sqrt{B^2 - 4CW4}}{2W4}$$

11-12d

The remaining mole numbers, $N(8)$, $N(9)$, and $N(10)$, are calculated from Eqs. 11-11h, 11-11i, and 11-11g.

This completes one iteration. By keeping track of the corresponding mole numbers computed in each previous iteration, we halt the iteration when the change in mole fraction values between successive iterations becomes as small as we like. When the mole fraction change for any specie is outside the allowable error, $N(1)$ is changed

$$N(1) = N(1) \pm \frac{AC}{2^{J+1}}$$

11-13

and we need to know which sign is appropriate. That is decided by comparing

$$W6' = \frac{N(10)}{N(8)\sqrt{N(3)}}$$

with $W6$. But we need also to avoid negative $N(10)$, and the quantity C must be negative. The logic for this program is obtained by differentiating Eqs. 11-11, treating $K3$ and $W1, \ldots, W6$ as constants, and writing each differential in terms of $dN(1)$. The results can be stated as follows:

If C is positive, increase $N(1)$;
If $N(10)$ is negative, decrease $N(1)$;
If $W6'$ is greater than $W6$, increase $N(1)$;
If $W6'$ is less than $W6$, decrease $N(1)$.

Each iteration on $N(1)$ begins with $J = 1$ in Eq. 11-13 and proceeds through a chain of calculations to derive values for the remaining nine mole numbers. If in the course of this set of calculations C shows up positive, $N(1)$ is increased, or if $N(10)$ shows up negative, $N(1)$ is decreased, the iteration is abandoned, J is indexed by unity, and the routine returns to Eq. 11-11a.

When the abundance of oxygen atoms, AO, is large compared to AH and AC, the program may not converge. The reason lies in the essential feature of the half-interval

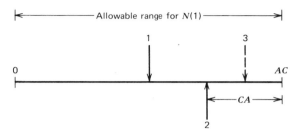

Figure 11.2 Iterations start at point 1, midrange, with $J = 1$. If J reaches 20, as at point 2, but C (Eq. 11–12b) is still positive, reset $J = 1$ and restart the search at point 3, a new midrange point.

search; the program closes in on the solution by halving the increment in $N(1)$ between successive iterations. As J in Eq. 11-13 becomes large, the change in $N(1)$ is correspondingly small. For example, with $J = 20$

$$\frac{1}{2^{21}} = 0.0000004$$

Since computers ordinarily work with seven significant figures, a point can be reached with increasing J at which the half-interval search ceases to be effective; with further increase in J, the values of the mole numbers will not change. There are two ways to fix this situation.

One way is to take the trouble to write the program in double precision, which doubles the number of significant figures. The other method is described in graphic form in Fig. 11.2. As shown, if J reaches, say, 20, and C is positive and $N(1)^*$ is the current value for $N(1)$, interrupt the iteration, introduce the quantity CA

$$CA = AC - N(1)^*$$

set

$$N(1) = AC - \frac{CA}{2}$$

reset $J = 1$, and continue, replacing the recursion expression Eq. 11-13 with

$$N(1) = N(1) \pm \frac{CA}{2^{J+1}}$$

Essentially, this redefines the allowable range for $N(1)$, and it can be done as many times as may be required, which will depend on the relative values of the abundancies AO, AC, and AH.

As in the previous program discussions,

1. The program will operate satisfactorily for $AN = 0$.
2. The program will not operate satisfactorily if $AO = 0$ or $AH = 0$ or $AC = 0$.
3. AO must be larger than AC.

Figure 11.3 is one illustration of the sort of results obtained from chemical equilibrium composition calculations. It reveals the profound influence of temperature. For the example drawn in that figure, $P = 1$ atm, and the ratios of the abundancies are

$$AC{:}AH{:}AO{:}AN = 10{:}22{:}40{:}0$$

which might, for example, represent the combustion products produced from a reactant mixture of a hydrocarbon or alcohol with oxygen. Treating the abundancies

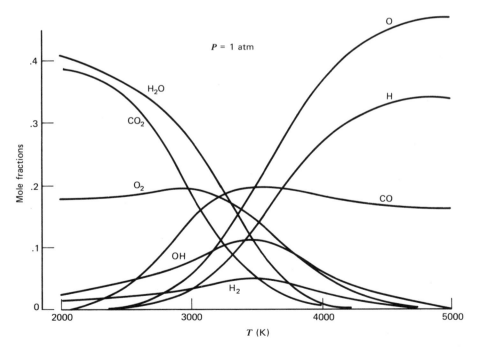

Figure 11.3 Equilibrium composition, showing the influence of temperature, for $AC/AH/AO/AN = 10/22/40/0$.

in units of moles, for example, below 2000 K the mixture consists of

$$10CO_2 + 11H_2O + 4.5O_2$$

but when heated to 5000 K, at constant pressure, the mixture will have changed to

$$10CO + 22H + 30O$$

Note that the mole fractions for H_2O and CO_2, the large molecules, decrease steadily with increasing temperature, while the mole fractions for the atoms O and H behave oppositely. The four diatomic molecules, O_2, H_2, CO, and OH all reach maxima at some temperature, so it is not generally known whether they will increase or decrease with a temperature change of small dimension. Note how rapidly the composition is changing in the region of 3500 K.

The example chosen for Fig. 11.3 involves quite dramatic changes in composition. These would be greatly suppressed if the oxygen were accompanied by the normal amount of nitrogen in air. Introducing what amounts to almost an inert gas will impede dissociation.

The shapes of the mole fraction curves are affected by the abundancy ratios and by the prevailing pressure. With a working program in hand, these can be explored and examined. It is advantageous to carry out this sort of exercise, as it enhances one's intuitions about the behaviour of ideal gas mixtures that are constrained to a condition of chemical equilibrium.

11.5 FURTHER EXTENSION, TO 12 OR MORE SPECIES

To the 10 species included in the solution described in Section 11.4 may be added atomic nitrogen, N, and solid carbon, C. The solution requires two more equations, which are mass action laws for

$$2N \rightleftharpoons N_2, \qquad K7$$
$$C + 2O \rightleftharpoons CO_2, \qquad K8$$

Also, a term for the N and C mole numbers is added to the conservation of matter equations. For this system, the technique of the half-interval search may be employed. One detail should be explained. Since the carbon is assumed to be solid, the mass action equation for the formation of CO_2 takes the form

$$K8 = \frac{N_{CO_2}}{N_O^2} \frac{NT}{P}$$

The entropy of a solid is assumed to be a function of temperature only, so the partial pressure attributable to solid carbon is zero.

The species specified in the mixture may be further extended by including various simple hydrocarbons with the form C_xH_y. However, the system of equations now takes on an entirely different character. The solution through a chain of calculations, pivoting about a single species, is no longer possible, and the mathematics makes a large jump to a level of vastly increased complexity.

11.6 SUMMARY

The composition of a system of ideal gases assumed to be in a state of chemical equilibrium is obtained by solving a set of awkward, nonlinear simultaneous equations, equal in number to the number of species in the mixture. Which species should be taken into account is entirely a matter of choice. Clearly, the major species should be included, and the minor species can be neglected. Unfortunately, it is seldom easy, at the outset, to predict with assurance which are major and which are minor. A glance at Fig. 11.3 will reinforce this point. In order to find out how plentiful a species may be in the mixture, that species must be included in the solution.

It is pointless to suppose that the system of equations can be effectively simplified, thereby bringing the algebra into more manageable form by some advantageous procedure, such as linear approximation. To do so merely defeats the entire purpose of the exercise. The only way to simplify is by reducing the number of species, leaving out those present in small amounts.

The character of the equilibrium composition is affected by the ratio of the abundancies of the elements present and by temperature and pressure. One of the difficult aspects of the numerical treatment of the system is the very wide range of values these controlling parameters can assume. This is particularly true for the reaction constants, as a glance at Table E.5 will demonstrate.

The general mathematical structure of the system of simultaneous equations has attracted a great deal of attention, mainly as a result of the advent of large memory, high-speed computers. A comprehensive survey of the work done during the period from 1950 to 1970 is contained in *The Computation of Chemical Equilibria*, van Zeggeren and Storey (Cambridge University Press, 1970). It contains about 200 references to work done during that period on a variety of systems, as well as a variety of methods for solution. The reader may find the book difficult because the mathematics is tightly organized and in places the discussions are brief. Furthermore, the book is concerned with the solutions for systems containing large numbers of species, say 30 or 40, which are vastly more complicated than the simple, restricted systems discussed here. The book is an excellent starting place for the reader who wishes to go on to further study.

For the reader who wishes a more extended program, going beyond 10 species, the National Aeronautics and Space Administration has published such a program, identified by NASA SP-273, entitled, "Computer Program for Calculation of Chemical Equilibrium Compositions, Rocket Performance, Incident and Reflected Shocks and Chapman-Jouget Detonations," by Sanford Gordon and Bonnie J. McBride (NASA

Lewis Research Center). The program, developed over a 20-year period, produces equilibrium composition by minimizing free energy, using a sophisticated Newton-Raphson iteration. Up to 150 species (molecules, atoms, and ions), including condensed phase (solid), may be present. An enormous range of reactants is possible. The program computes conditions following constant volume and constant pressure combustion, as well as rocket performance for exotic fuel/oxidizer combinations. A description of the program can be obtained from National Technical Information Service, Springfield, Virginia, 22151. A copy of the program itself, on magnetic tape, can be obtained from the authors.

PROBLEMS

11.1   Compose a computer program for calculation of the equilibrium composition of a system containing C, H, O, and N atoms, in which there are six assignable parameters, namely the four abundancies, the total pressure, and the temperature, and in which the species present in the mixture are restricted to CO, CO_2, H_2, H_2O, N_2, and O_2.

11.2   Modify the program of Problem 11.1 so that it applies to the same system, but with volume rather than total pressure as one of the assignable parameters.

11.3   Extend the programs in Problems 11.1 and 11.2 so that they apply to systems containing 10 species, namely CO, CO_2, H, H_2, O, O_2, H_2O, N_2, NO, and OH.
These programs may be checked against the results shown in Fig. 11.3.

For any computer program designed to compute equilibrium compositions, it is a good idea to check for convergence for the following conditions:
   (a)   Extreme temperatures, say, 5500 K and 1800 K,
   (b)   Extreme oxygen abundancies, very little and large excess,
   (c)   Stoichiometric oxygen abundance.

12
CHEMICAL EQUILIBRIUM IN H/O/N SYSTEMS

12.1 INTRODUCTION

The topic in this chapter is a continuation of the discussion of Chapter 11. Calculation of equilibrium mixture compositions in systems containing only H, O, and N atoms follows methods that closely parallel those employed in mixtures that contain C, H, O, and N atoms. Two sets of calculations are described: the H/O system with six species, and the H/O/N system with eight species. When carbon atoms are removed from the mixture, the calculation follows a somewhat different path, one that deserves to be examined as a separate exercise. The chief difference is found to be more manipulation of the governing equations, with a subtle difference in the checks for errors.

12.2 THE H/O SYSTEM WITH SIX SPECIES

The calculation of equilibrium mixture composition with six species will be examined. The same calculation with fewer than six species is readily evident. The mixture is identified as

$$N(1)H_2O + N(2)H_2 + N(3)O_2 + N(4)H + N(5)O + N(6)OH \qquad 12\text{-}1$$

with chemical equilibrium maintained by the following set of four independent reactions:

$$H_2 + \tfrac{1}{2}O_2 \rightleftharpoons H_2O \qquad \text{reaction constant } K1$$
$$2H \rightleftharpoons H_2 \qquad \text{reaction constant } K2$$
$$H + O \rightleftharpoons OH \qquad \text{reaction constant } K3$$
$$2O \rightleftharpoons O_2 \qquad \text{reaction constant } K4$$

With *AO* and *AH* denoting the abundance of oxygen and hydrogen atoms, respectively, *P* denoting total pressure, and *NT* denoting total mole number, the six equations from which the six mole numbers can be derived are as follows:

$$AH = 2[N(1) + N(2)] + N(4) + N(6)$$ 12-2

$$AO = 2N(3) + N(1) + N(5) + N(6)$$ 12-3

$$W1 = K1\sqrt{\frac{P}{NT}} = \frac{N(1)}{N(2)\sqrt{N(3)}}$$ 12-4

$$W2 = K2\frac{P}{NT} = \frac{N(2)}{N(4)^2}$$ 12-5

$$W3 = K3\frac{P}{NT} = \frac{N(6)}{N(4)N(5)}$$ 12-6

$$W4 = K4\frac{P}{NT} = \frac{N(3)}{N(5)^2}$$ 12-7

The known value of *T* fixes the values of the four reaction constants, using Eq. 10-16, with *a*, *b*, *c*, and *d* coefficients listed in Table C.2. The solution follows the half-interval search strategy, which can be set in motion after some initial manipulation.

First, solve Eq. 12-4 for *N*(1).

Then, solve Eq. 12-5 for *N*(4),

solve Eq. 12-6 for *N*(6),

solve Eq. 12-7 for *N*(5),

and eliminate these four mole numbers from Eq. 12-3, with the result

$$\sqrt{N(3)} = \frac{-B + \sqrt{B^2 + 8A\overline{O}}}{4}$$ 12-8

where

$$B = \frac{1}{\sqrt{W4}} + W1\,N(2) + W7\sqrt{N(2)}, \qquad W7 = \frac{W3}{\sqrt{W2W4}}$$ 12-9

Then

1. If *V* the volume and *T* are known, *P*/*NT* is known

$$\frac{P}{NT} = \frac{RT}{V}$$

If P and T are given, assume a reasonable value for NT, to be checked later in the solution against a calculated value, until assumed and calculated values agree within some assigned error.

2. Set $J = 1$ and start the half-interval search with $N(2)$ in the middle of its allowable range,

$$N(2) = \frac{AH}{4}$$

3. Compute $N(3)$ from Eq. 12-8 and the remaining four mole numbers from the mass action equations, Eqs. 12-4 through 12-7.

4. Compare the mole numbers $N(I)$ with $Z(I)$, the latter representing corresponding mole numbers from the previous iteration. (For the first pass, $Z(I)$ may be assigned arbitrary values.)

5. If

$$\frac{|N(I) - Z(I)|}{NT} \leqslant \text{error}, \qquad I = 1, 6$$

where *error* is a preassigned limit, then the calculation is complete if V and T are known. If P and T are known, compare NT with NT'

$$NT' = \Sigma N(I)$$

and if the difference is within tolerable error, which could be 1% or 2%, say, the solution is complete. Otherwise, set $NT = NT'$, and return to step 2.

6. If the difference between any pair of consecutive mole numbers in step 5 exceeds the *error*, then $N(2)$ is changed,

$$N(2) = N(2) \pm \frac{AH}{2^{J+2}} \qquad \text{12-10}$$

and the proper choice of sign is dictated by the error in the unused equation, Eq. 12-2, which can be written now as

$$Q = 2[1 + W1\sqrt{N(3)}]N(2) + \left[\frac{1}{\sqrt{W2}} + W7\sqrt{N(3)}\right]\sqrt{N(2)} - AH = 0 \qquad \text{12-11}$$

Obviously, we need to know the algebraic sign of the differential quotient

$$\frac{dQ}{dN(2)}$$

which is to be found by differentiating Eq. 12-11 and eliminating all mole number differentials, using the mass action equations 12-4 through 12-7, in terms of $dN(2)$. This exercise is left to the reader.

One further comment: At low temperature, the quantity B, which occurs in Eq. 12-8 and is defined in Eq. 12-9, may become large compared to $8AO$. When this happens, $N(3)$ is calculated from Eq. 12-8 as the difference between two large and approximately equal numbers, with a result that may be unreliable. When B becomes large, therefore, expand the right-hand side of Eq. 12-8 and calculate $N(3)$ by

$$\sqrt{N(3)} = \frac{B}{4}\left[\frac{1}{2}Y - \frac{1}{8}Y^2 + \frac{1}{16}Y^3 - \cdots\right] \qquad \text{12-12}$$

where

$$Y = \frac{8AO}{B^2}$$

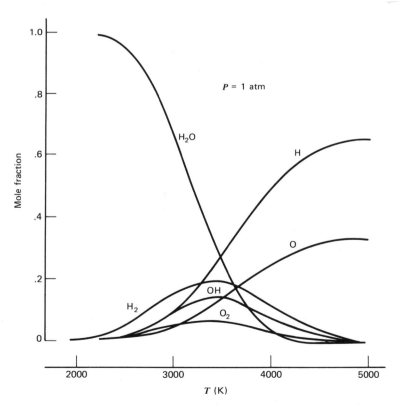

Figure 12.1 Equilibrium composition, *AO/AH* = 1/2.

It is left to the reader to determine at what B value Eq. 12-8 is to be abandoned in favor of Eq. 12-12.

Figure 12.1 illustrates the influence of temperature, at $P = 1$ atm, on an H/O system with $AO/AH = 1/2$. Note the similarity with Fig. 11.3. At low temperature, only the triatomic molecule H_2O is present, while at high temperature, only the H and O atoms are present. At 3500 K, where the composition is changing rapidly, atoms and diatomic and triatomic molecules are present in more nearly equal amounts.

12.3 THE H/O/N SYSTEM WITH EIGHT SPECIES

Before developing the solution with eight species, we might note a possible simplification. When the mixture of gases for which we seek an equilibrium composition results from the combustion of hydrogen and *air*, a product mixture restricted to four species,

$$H_2, O_2, H_2O, \text{ and } N_2 \qquad 12\text{-}13$$

in chemical equilibrium through the single reaction

$$H_2 + \tfrac{1}{2}O_2 \Longleftrightarrow H_2O$$

may, for thermodynamic purposes at least, be a satisfactory approximation to the real state of affairs. The reason for this is twofold. With air supplying the oxygen for combustion, a significant fraction of the energy released will be absorbed by N_2, simply because there is an abundance of that species. And further, if the prevailing pressure is large, that condition will suppress the tendencies of triatomic molecules to dissociate to diatomic molecules which then dissociate to atoms. Referring to Fig. 12.1, those curves suggest that if 12-13 is the assumed species group in the product mixture, and the temperature is 2500 K or less, or perhaps 3000 K or less, then 12-13 may be assumed satisfactory.

Turning to the solution for an equilibrium mixture containing eight species,

$$N(1)H_2O + N(2)H_2 + N(3)O_2 + N(4)H + N(5)O + N(6)OH + N(7)N_2 + N(8)NO$$

$$12\text{-}14$$

eight equations are required. Three are supplied by the conservation equations

$$AN = 2N(7) + N(8) \qquad 12\text{-}15$$

$$AH = 2[N(1) + N(2)] + N(4) + N(6) \qquad 12\text{-}16$$

$$AO = 2N(3) + N(1) + N(5) + N(6) + N(8) \qquad 12\text{-}17$$

The mixture is in chemical equilibrium through the five reactions

$$H_2 + \tfrac{1}{2}O_2 \; \rightleftharpoons \; H_2O, \qquad \text{reaction constant } K1$$
$$2H \; \rightleftharpoons \; H_2 \qquad\qquad \text{reaction constant } K2$$
$$H + O \; \rightleftharpoons \; OH \qquad\quad \text{reaction constant } K3$$
$$2O \; \rightleftharpoons \; O_2 \qquad\qquad \text{reaction constant } K4$$
$$O_2 + N_2 \; \rightleftharpoons \; 2\,NO \qquad \text{reaction constant } K5$$

which lead to five mass action equations, Eqs. 12-4 through 12-7, and

$$K5 = \frac{N(8)^2}{N(3)N(7)} \tag{12-18}$$

Solve the mass action equations as follows:

$$N(1) = W1\,N(2)\sqrt{N(3)}$$

$$N(4) = \sqrt{\frac{N(2)}{W2}}$$

$$N(5) = \sqrt{\frac{N(3)}{W4}} \tag{12-19}$$

$$N(6) = W7\sqrt{N(2)N(3)}$$

$$N(8) = \sqrt{K5\,N(3)N(7)}$$

where $W7$ is defined in Eq. 12-9. After substitution for these mole numbers in Eqs. 12-15 and 12-16, we have

$$AN = 2N(7) + \sqrt{K5\,N(3)N(7)}$$

$$AH = 2[1 + W1\sqrt{N(3)}]\,N(2) + \sqrt{\frac{N(2)}{W2}} + W7\sqrt{N(2)N(3)} \tag{12-20}$$

A half-interval search can now proceed, using $N(3)$ as the pivotal variable, starting in the middle of the allowable range

$$N(3) = \frac{AO}{4}$$

The remaining seven mole numbers can now be calculated from Eqs. 12-19 and 12-20. There remains Eq. 12-17 to be satisfied. Writing that equation in the form

$$Q = 2N(3) + N(1) + N(5) + N(6) + N(8) - AO \qquad 12\text{-}21$$

the differential quotient

$$\frac{dQ}{dN(3)}$$

informs the program how the algebraic sign of Q determines the appropriate sign in the expression

$$N(3) = N(3) \pm \frac{AO}{2^{J+2}}$$

in which $J = 1$ is the proper index for J at the start of the search. The differentiation of Eqs. 12-19, 12-20, and 12-21 is left to the reader. It is cumbersome but straightforward.

Question: Is the program for the H/O system described in Section 12.2 a special case of the program for the H/O/N system discussed here? Or to put the question more directly, will the program of this section operate satisfactorily for $AN = 0$?

12.4 SUMMARY

The programs for equilibrium composition calculations that have been discussed in this chapter and in Chapter 11 employ the strategy of the half-interval search. There is a subtle difference between the two, which ought not to be passed over.

In Chapter 11, dealing with C/H/O/N systems, the order of calculations was such that the conservation of matter equations were satisfied; the error, or the allowable tolerance, in the solution occcurred in one of the mass action equations. In this chapter, dealing with H/O/N systems, the order of calculation places the error in one of the conservation of matter equations. In a certain sense, conservation of matter is more "fundamental" than mass action. Losing track of matter is less to be desired than not satisfying maximum entropy exactly.

It is sound practice, then, to check out the degree of error in computer programs that throw that error into a conservation equation. In the last program of this chapter, for example, when the changes in mole fractions between successive iterations have been reduced to an acceptable level, return to Eq. 12-17 or Eq. 12-21 (they are identical) and compare the given value of AO with

$$2N(3) + N(1) + N(5) + N(6) + N(8)$$

The two should agree closely, say within 1% or less. If not, then tighter tolerance is needed on the mole fraction agreement, and further iterations are required.

The last comment has to do with the selection of the reversible chemical reactions through which chemical equilibrium is maintained. In mixtures that permit more than one reaction, there is usually an abundance of reactions, many more than are needed to fill out the required number of equations for the solution. How should one select from this overabundance? Are some sets of reactions better than others or easier to deal with? From the mathematical point of view, all sets of reactions must lead to the same equilibrium composition. The question of "better" and "easier" is settled by the programmer's habits or preferences. In the case of programs discussed in this and in the preceding chapter, those reactions were chosen for which a, b, c, d values in Table C.2 were available. The decisions were based on convenience. Co-efficients for other reactions can, of course, be produced, as explained in Appendix C, if they are needed or preferred.

PROBLEMS

12.1   Compose a computer program for calculation of the equilibrium composition of a system containing H and O atoms in which there are four assignable parameters, namely the two abundancies, the total pressure, and the temperature, and in which the species present are restricted to H_2O, H_2, O_2, H, O, and OH. The program may be checked against Fig. 12.1.

12.2   Modify the program of Problem 12.1 so that it applies to the same system, but with volume rather than total pressure as one of the assignable parameters.

12.3   Compose a computer program for calculation of the equilibrium composition of a system containing H, O, and N atoms in which there are five assignable parameters, namely the two abundancies, the total pressure, and the temperature, and and in which the species present in the mixture are restricted to H_2O, H_2, O_2, H, O, OH, N_2, and NO. Then modify the program so that volume rather than total pressure is assignable.

For any computer program designed to calculate equilibrium compositions, it is a good idea to check for convergence under the following conditions:
   (a) Extreme temperatures, say, 5500 K and 1800 K,
   (b) Extreme oxygen abundancies, very little and large excess,
   (c) Stoichiometric oxygen abundance.

13

THE ADIABATIC FLAME TEMPERATURE AND ISENTROPIC EXPANSIONS, FOR SYSTEMS IN CHEMICAL EQUILIBRIUM

13.1 INTRODUCTION

The analysis of all combustion engine cycles requires calculation of the adiabatic flame temperature following combustion and calculation of the final state reached after an isentropic expansion. In Chapter 2 we saw how the flame temperature computation can be carried out for a variety of conditions, all under the assumption of complete combustion, that is, with the product mixture composition known at the outset. In Chapter 3 we learned how to treat isentropic expansions, under the assumption of unchanging composition in the gaseous mixture. Equipped with the techniques for computing equilibrium compositions discussed in Chapters 11 and 12, we turn now to examine flame temperature and expansion computations for systems that are in chemical equilibrium. Two methods are described: the Newton-Raphson iteration and the half-interval search. One of the objectives of this chapter is to reveal the difficulties that lurk within the Newton-Raphson method and render that method unsuited for systems that contain large numbers of species in the product gas mixture.

13.2 NEWTON-RAPHSON ITERATION

The nature of the problem encountered with a Newton-Raphson iterative solution is easily demonstrated with an elementary system which reduces the algebra to manageable form.

We wish to calculate the adiabatic flame temperature following combustion of a gaseous mixture

$$H_2 + \tfrac{1}{2}O_2$$

initially at 298 K, when the combustion takes place at constant pressure P. We shall examine the mechanics of the solution for various P values. The equilibrium product mixture is restricted to

$$n_{H_2O} H_2O + n_{H_2} H_2 + n_{O_2} O_2$$

and equilibrium is maintained by the single reaction

$$H_2 + \tfrac{1}{2}O_2 \rightleftharpoons H_2O \qquad\qquad 13\text{-}1$$

The three equations for the mole numbers are

$$
\begin{aligned}
n_{H_2} + n_{H_2O} &= 1 \\
n_{H_2O} + 2n_{O_2} &= 1
\end{aligned}
\qquad\qquad 13\text{-}2
$$

$$K_p = \frac{n_{H_2O}}{n_{H_2}\sqrt{n_{O_2}}} \sqrt{\frac{n_{H_2O} + n_{H_2} + n_{O_2}}{P}} \qquad (P \text{ in atm})$$

The energy equation for T the flame temperature is

$$H_r(298) = H_p(T) \qquad\qquad 13\text{-}3$$

subscripts r and p denoting reactants and products. Let

$$F(T) = H_p(T) - H_r(298) \qquad\qquad 13\text{-}4$$

The solution for $F(T) = 0$ is obtained by repeated application of the recursion expression

$$TNEW = T - \frac{F(T)}{C_p^*} \qquad\qquad 13\text{-}5$$

where

$$C_p^* = F'(T) = \frac{d}{dT} F(T) = \frac{d}{dT} \Sigma\, n_i h_i(T)$$

$$= \Sigma\, n_i \frac{dh_i}{dT} + \Sigma\, h_i \frac{dn_i}{dT}$$

$$C_p^* = C_p + \Sigma\, h_i \frac{dn_i}{dT} \qquad (P \text{ constant}) \qquad \text{13-6}$$

Here C_p is the conventional constant pressure heat capacity of the mixture (kJ/K). When the composition of the product mixture is assumed fixed, then all $dn_i = 0$, and $C_p^* = Cp$, since the second term in Eq. 13-6 drops out. When the product mixture is in equilibrium, the mole numbers are functions of P and T and the second term in Eq. 13-6 will be nonzero. To evaluate that term, differentiate Eqs. 13-2, eliminating dn_{H_2} and dn_{O_2}. The result can be written

$$\frac{dK_p}{K_p} = \frac{dn_{H_2O}}{\Psi N} \qquad \text{13-7}$$

where N is the total mole number and

$$\frac{1}{\Psi} = \frac{1}{x_{H_2O}} + \frac{1}{x_{H_2}} + \frac{1}{4x_{O_2}} - \frac{1}{4} \qquad \text{13-8}$$

a quantity introduced in discussing LeChatelier's equations. Making use of van't Hoff's equation

$$\frac{dK_p}{K_p} = \frac{\Delta H}{RT^2} dT$$

where

$$\Delta H = (h_{H_2O} - h_{H_2} - \tfrac{1}{2}h_{O_2})_T \qquad \text{13-9}$$

Eq. 13-7 becomes

$$\frac{dn_{H_2O}}{dT} = \frac{N\Psi(\Delta H)}{RT^2} \qquad (P \text{ constant}) \qquad \text{13-10}$$

Since conservation of matter requires

$$dn_{H_2O} = -dn_{H_2} = -\tfrac{1}{2}dn_{O_2}$$

Eq. 13-6 can be written

$$C_p^* = C_p + (h_{H_2O} - h_{H_2} - \tfrac{1}{2}h_{O_2})\frac{dn_{H_2O}}{dT}$$

$$= C_p + (\Delta H)\frac{dn_{H_2O}}{dT}$$

and with Eq. 13-10, we have the final form

$$C_p^* = C_p + \frac{N\Psi(\Delta H)^2}{RT^2} \qquad\qquad\text{13-11}$$

Note that ΔH occurs squared in Eq. 13-11. Had we written

$$H_2O \;\rightleftharpoons\; H_2 + \tfrac{1}{2}O_2 \qquad\qquad\text{13-12}$$

instead of Eq. 13-1, the sign of ΔH would be reversed. Since C_p^* is a physical property of the mixture, it must be independent of the way we choose to write the reaction that maintains equilibrium.

For various (P, T) values, the mole numbers, $F(T)$, C_p and C_p^* may be computed. Figure 13.1 shows some results for two pressures, $P = 1$ atm and $P = \infty$. In the latter case, Eqs. 13-2 require $n_{H_2} = n_{O_2} = 0$; that is, infinite pressure prohibits any dissociation and leads to the complete combustion mixture. Figure 13.1 is a graphic comparison of complete combustion and chemical equilibrium solutions. The flame temperatures, found for $F(T) = 0$, are 5200 K and 3600 K. For intermediate pressures, we can expect the flame temperatures to lie between these extremes.

Note that C_p^* reaches a maximum with $P = 1$ atm. At the maximum, $C_p = 54$ kJ/K and $C_p^* = 167$ kJ/K. This means that two-thirds of the energy added to the mixture, at constant pressure, will be deployed in dissociating H_2O molecules, while the remaining one-third will increase mixture temperature. Note, also, that C_p^* is everywhere greater than C_p. This result can be shown to be true mathematically (see Appendix D). From physical argument, it is clear that these two heat capacities must be related in this way. Adding energy to a mixture that contains large particles which can dissociate requires more energy, for the same temperature change, than does the same mixture which is in some way prohibited from any chemical reaction.

Because C_p^* is the slope of the function $F(T)$, Eq. 13-6, and because C_p^* exhibits a maximum, $F(T)$ will be S-shaped plotted against T. Figure 13.2 shows an exaggerated

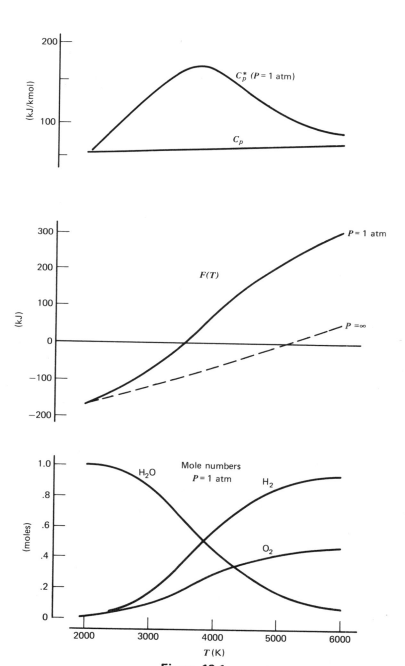

Figure 13.1

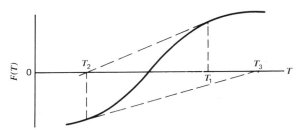

Figure 13.2 The problem of divergence when the Newton-Raphson iteration is applied to an S-shaped curve.

case simply for clarity. (For $P = 1$ atm, the S-shaped character of $F(T)$ can be seen in Fig. 13.1) Now when the iteration routine is applied, Eq. 13-5, with T_1 denoting the first approximation and T_2 and T_3 the successive approximations, the solution will not converge, as shown in the figure, because T_3 is greater than T_1. Successful convergence requires a suitable T_1. To be sure, finding one is not a difficult assignment, and for this special case with a single reaction maintaining chemical equilibrium, the Newton-Raphson iteration could be carried through easily.

   The same cannot be said for multireaction systems that are characterized by more complicated mathematical structure, which increases in its complexity as the number of reactions increases. In particular, the expression for C_p^* cannot be reduced to anything approaching the neat conciseness of Eq. 13-11. The recursion equation, Eq. 13-5, applies to multireaction mixtures, but we are now forced to compute C_p^* from its primitive definition

$$C_p^* = \frac{H_p(T + \Delta T) - H_p(T)}{\Delta T} \qquad 13\text{-}13$$

which requires small ΔT for accuracy. But with small ΔT, extreme accuracy is required in the equilibrium composition computations at T and $T + \Delta T$, which furnish the mole numbers that occur in $H_p(T)$ and $H_p(T + \Delta T)$.

   Turning to isentropic expansions from known state (P_1, T_1) to a known final pressure, P_2, the corresponding final temperature T, is the solution to

$$F(t) = S(P_2, T) - S(P_1, T_1) = 0 \qquad 13\text{-}14$$

where $S(P, T)$ represents total entropy of the mixture. As developed in Eq. 13-10, which applied to mixtures with fixed composition, the recursion expression for the Newton-Raphson solution of Eq. 13-14 for the case of equilibrium mixtures is

$$\text{TNEW} = T \left[1 - \frac{F(T)}{C_p^*(T)} \right] \qquad 13\text{-}15$$

and there is no escape from C_p^* computations.

For multireaction systems, it is best to abandon the Newton-Raphson iteration in favor of the half-interval search, which involves only the enthalpy, internal energy, and entropy of equilibrium mixtures.

13.3 HALF-INTERVAL SEARCH

The anatomy of the half-interval search routine is discussed in some detail in Appendix B and is described in a flow chart, Fig. B.2. That chart may be directly applied to flame temperature determinations,

$$F(T) = H_p(T) - H_r(T_r)$$
$$F(T) = U_p(T) - U_r(T_r)$$

because $F(T)$ increases with T, and $F(T_r)$ is negative. For isentropic expansions,

$$F(T) = S(P_2, T) - S(P_1, T_1), \qquad P_1 > P_2$$
$$F(T) = S(V_2, T) - S(V_1, T_1), \qquad V_2 > V_1$$

$F(T)$ increases with T, and $F(T_1)$ is positive, which requires slight modification of the flow chart.

In two places the flow chart contains the block figure

$$\boxed{\text{Is } F(T) > 0 \quad ?}$$

The reader should be aware of the programming required to answer that question. When that block is expanded to reveal the calculations that are hidden from casual glance, the result is

---

1. For the given abundancies, and for the values of T and P (or V) currently applicable, compute the mole numbers in a equilibrium mixture.
2. Compute $H_p(T)$ or $U_p(T)$ or $S(P_2, T)$ or $S(V_2, T)$ as required.
3. Compute $F(T)$
4. Is $F(T) > 0$?

Since this block of calculations is called for more than once, it is convenient to arrange the equilibrium composition calculation in a subroutine that can be called at any point in the program. Similarly, program construction is vastly simplified when function programs for computation of H, U, and S for the mixture are written.

13.4 SUMMARY

The objective in this chapter has been to outline some of the pitfalls that surround the use of the Newton-Raphson iteration in treating mixtures of ideal gases in equilibrium and to recommend adoption of the half-interval search as a reliable alternative. The generally slower convergence of the latter, as compared to the former, is more than made up for by its simplicity and its assured march to a solution.

Further inquiry into the behavior of gas mixtures raises important questions that go beyond the limited discussion of this chapter. For example, when a set of reactants burns, under whatever conditions, will the product mixture reach chemical equilibrium? If it does not, what composition will be reached? When a mixture in equilibrium is disturbed, will equilibrium be maintained? If not, how will the composition change? This preoccupation with the composition of the mixture is readily understandable from a glance at the h and ϕ tables in Appendix E; the values of these functions, particularly h, vary substantially among those species that *could be* present in the mixture. The composition of the mixture influences the values of U, H, and S and, through them, derived thermodynamic quantities such as the flame temperature, the work transfer during expansion, and the velocity reached during expansion.

Answers to these questions go beyond the reach of thermodynamics, which is confined to time-independent states that can be described by the application of the first and second laws and the law of conservation of matter. The reactions that establish and maintain chemical equilibrium proceed at finite rates that are strongly influenced by temperature and, to a lesser degree, by pressure. Even at elevated temperature where the rates are high, the conversion of a finite quantity of A to $B + C$ requires a finite amount of time. And there may be a host of intermediate chemical species and reactions that are essential to that conversion.

When questions are asked in a context from which time is removed, we can answer with more, but not complete, assurance. When a well-mixed set of reactants burns at constant volume, the bulk of the products are undoubtedly in chemical equilibrium. There is an accumulating body of experimental evidence which indicates that the reactants near the wall reach a nonequilibrium state in which partially decomposed fuel molecules are present. The advancing flame is quenched as it approaches the wall. These findings lead one to conclude that the amount of surface area in the enclosure will exert a controlling effect on the quantity of nonequilibrium mixture present.

When time enters the process, the questions are more difficult to answer. How does the mixture in the chamber of an engine, rotary or reciprocating piston, behave during expansion? The rate of expansion is a reflection of engine speed and can vary over wide limits. Likewise, what happens to a mixture in equilibrium in a rocket motor combustion chamber as it passes along the converging-diverging nozzle?

If our treatment of combustion is confined to information that can be supplied by thermodynamics, we have only two choices for the composition of the products of combustion: either the mixture is in chemical equilibrium or the composition is derived without regard for the second law, which we call complete combustion. Undoubtedly, there are many circumstances in which the composition falls somewhere between the two. The rules and experimental data required to determine the real composition are emerging slowly, for the study of reaction kinetics at high temperature is a complex tangle that is not yet well understood.

14

THE EFFECTS OF CHEMICAL EQUILIBRIUM ON PERFORMANCE CALCULATIONS

Performance calculations are generally brief and straightforward when the composition of the exhaust products is based on the assumption of complete combustion. The assumption of chemical equilibrium extends and complicates these calculations. It is natural to inquire whether the calculated performance of devices that derive their motive power from combustion is affected substantially by including chemical equilibrium as a basic assumption in the analysis. The answers to this question are the subject of this chapter.

By way of introduction, Fig. 14.1 shows the adiabatic flame temperature as a function of equivalence ratio for constant volume combustion of $C_{10}H_{22}$ with air and with gaseous oxygen. The calculations have been carried out for three systems:

1. Complete combustion (H_2O, CO_2, CO, N_2 for rich mixtures; H_2O, CO_2, O_2, N_2 for lean mixtures)
2. Chemical equilibrium (H_2O, CO_2, CO, H_2, O_2, N_2)
3. Chemical equilibrium (H_2O, CO_2, CO, H_2, O_2, NO, O, H, OH)

14-1

In the case of $C_{10}H_{22}$/air mixtures, the discontinuity at chemically correct condition, which is the result of the abrupt change in composition with system 1, is smoothed out with systems 2 and 3. The temperatures predicted by the three systems differ by less than 100 K. Consequently, for the initial conditions of 298 K and 1 atm, complete combustion is "good enough." Note that in the region of low equivalence ratios, systems 2 and 3 lead to flame temperatures slightly larger than those calculated with system 1. For the case of $C_{10}H_{22}$/O_2 mixtures, the temperature differences are enormous, and for systems 2 and 3, the flame temperature is virtually independent of equivalence ratio.

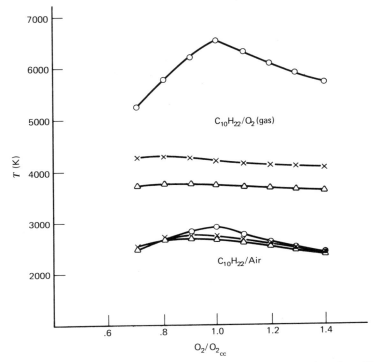

○ Complete combustion (CO, CO_2, H_2O, O_2, N_2)
✕ Chemical equilibrium (CO, CO_2, H_2O, H_2, O_2, N_2)
△ Chemical equilibrium (CO, CO_2, H_2O, H_2, O_2, N_2, NO, H, O, OH)

$C_{10}H_{22}/O_2$ (gas)

$C_{10}H_{22}/$Air

$O_2/O_{2_{cc}}$

Figure 14.1 Adiabatic flame temperature. Constant volume combustion, $T_r = 298$ K, $P_r = 1$ atm.

Figure 14.2 is a corresponding plot for constant pressure combustion. The pattern is similar to Fig. 14.1, and the temperatures are slightly lower. (Why?)

Because the heats of reaction *per unit mass* are practically independent of the composition and structure of hydrocarbon fuels (see Table E.4), Figs. 14.1 and 14.2 are representative of all hydrocarbons. The alcohols will produce essentially similar curves.

Performance calculations based on complete combustion are satisfactory for gas turbines, boilers, and furnaces. These devices use air to support combustion at constant pressure. A possible exception may be the afterburner section of a high compression aircraft gas turbine.

For furnaces fired with fuel and oxygen or with a special oxygen/nitrogen mixture, *design* would be based on flame temperature computations that include

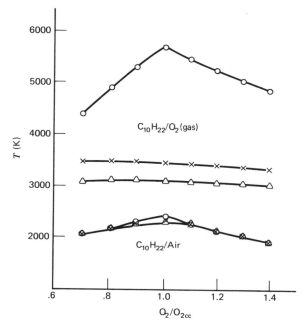

O Complete combustion (CO, CO$_2$, H$_2$O, O$_2$, N$_2$)
× Chemical equilibrium (CO, CO$_2$, H$_2$O, H$_2$, O$_2$, N$_2$)
△ Chemical equilibrium (CO, CO$_2$, H$_2$O, H$_2$, O$_2$, N$_2$, NO, H, O, OH)

Figure 14.2 Adiabatic flame temperature. Constant pressure combustion, T_r = 298 K, P = 1 atm.

equilibrium. Furnace *performance*, particularly the rate of heat loss, involves properties of the gases in the exhaust stack, which can be adequately determined on the basis of complete combustion theory. This procedure is based on the rule-of-thumb that dissociation may be ignored below 2000 K, provided pressure is of the order of 1 atm or more.

Figure 14.3 illustrates power output of a four-stroke Otto cycle engine drawn against equivalence ratio. These calculations are based on the sharp-cornered, ideal cycle. Complete combustion figures, for example, are taken from Table 5.5a. For systems 2 and 3 above, power decreases except in the extreme fuel-rich region. The most noticeable feature of Fig. 14.3 is the fact that maximum power with systems 2 and 3 occurs, not at chemically correct mixture ratio, but at fuel-rich mixtures, a fact readily confirmed by experiment.

In the case of rockets driven by fuel/oxygen reactant mixtures, chemical equilibrium exerts a profound effect because of the high flame temperature developed in

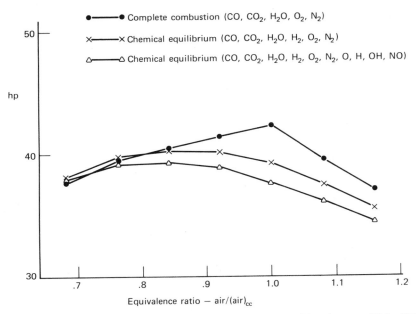

Figure 14.3 Four-stroke ideal Otto cycle. Full throttle. $P1 = 1$ atm; $TM = 300$ K; speed = 3000 rpm; $CR = 9{:}1$; VDISP = .000823 m^3.

the combustion chamber. Figure 14.4 illustrates the variation in adiabatic flame temperature for liquid H_2/liquid O_2 mixtures, plotted against equivalence ratio, for constant pressure combustion at 20 atm. The calculations were carried out for three product systems;

1. Complete combustion (H_2, H_2O for mixtures with excess H_2; and O_2, H_2O for mixtures with excess O_2)
2. Chemical equilibrium (H_2, O_2, H_2O)   14-2
3. Chemical equilibrium (H_2, O_2, H_2O, H, O, OH)

As in Figs. 14.1 and 14.2, the discontinuity in the complete combustion curve is removed when chemical equilibrium in introduced, and the temperatures for systems 2 and 3 are nearly constant over a considerable range of equivalence ratio.

Figure 14.5 illustrates the large differences in the predictions of rocket performance that result from different assumptions for the composition of the products of combustion. The performance parameter of interest is the altitude reached at the end of the coasting period. Gravitational acceleration is assumed constant during powered flight but is variable during the coasting period. The rocket is single stage, liquid H_2/liquid O_2, lift-off mass = 13 605 kg, propellant mass = 12 245 kg, nozzle throat

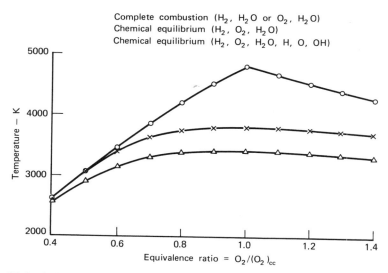

Figure 14.4 Adiabatic flame temperature. Constant pressure combustion. Liquid H_2/liquid O_2 P = 20 atm.

area = 0.056 m² combustion chamber pressure = 20 atm, and nozzle exit section pressure = 1 atm. The mass flow rated is derived from Eq. 7-28 (repeated here)

$$\dot{M} = \Gamma A t P c \sqrt{\frac{MW}{RTc}}$$ 14-3

which assumes constant composition (i.e., constant molecular weight MW) as combustion products flow from the combustion chamber to nozzle throat.

Although it is reasonable to assume that chemical equilibrium prevails in the combustion chamber, it is not clear how the gas mixture will behave as it passes along the nozzle. We could assume that expansion is so rapid that there is no time for the reactions necessary to maintain chemical equilibrium, so the composition remains fixed at combustion chamber values. This is referred to as "frozen flow." On the other hand, we can just as well assume the mixture remains in a state of chemical equilibrium during passage through the nozzle. This is referred to as "shifting equilibrium." The difference between the two assumptions is enormous, as Fig. 14.5 shows. Which is the correct assumption? Probably the best answer is, neither. More than likely, the actual combustion gas behavior lies somewhere between the two assumptions.

As pointed out in Chapter 10, the collision frequency in gaseous systems is of the order of a billion per second. Even if the passage time through the nozzle is only a millisecond, the number of collisions remains high at one million. But is that number high enough to maintain chemical equilibrium, that is, to distribute the total energy

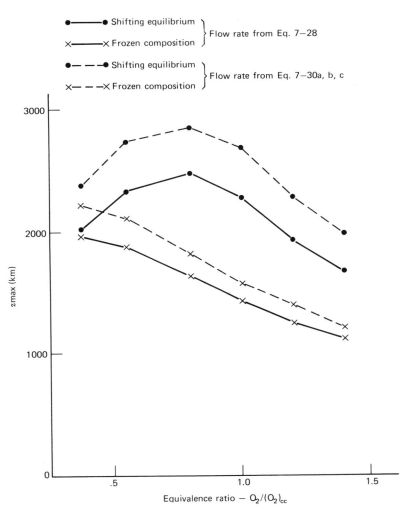

Figure 14.5 Liquid H_2/liquid O_2 rocket, fired vertically from earth. Mo = 13 605 kg; Mp = 12 245 kg (π = .9); At = .056 m^2, Pc = 20; Pe = 1 atm. Products of combustion; H_2, O_2, H_2O, O, H, OH. Constant g during powered flight, variable g during coasting.

continuously among the species and among the allowable forms of motion, translation, rotation, and vibration? Collisions distribute translational energy rapidly, but rotational and vibrational energy are distributed at slower rates and lag behind by what is known as "relaxation times." Rational calculation of the probable composition of the combustion gases as they pass through the nozzle is a formidable undertaking (quite beyond the aims of this book) and generally requires more information about the behavior of gases at high temperature than is usually available. The true performance of a rocket in simple vertical flight is to be ascertained by testing, not by calculation. For purposes of analysis, thermodynamics provides but two options, frozen flow or shifting equilibrium. There is no logic, within thermodynamics, for any other choice.

The formulation of the mass flow rate, M, in Eq. 14-3 is derived under the assumptions of constant composition and constant heat capacities. Since neither assumption is compatible with shifting equilibrium, Eq. 14-3 may introduce error. To calculate the mass flow rate with shifting equilibrium we must resort to the expression

$$\dot{M} = \rho_t A_t V_{a,t} \qquad\qquad 14\text{-}4$$

where $V_{a,t}$ is the acoustic velocity at the throat of the nozzle and is computed by observing the behavior of the gas mixture as it passes along the nozzle. In doing so, pressure, temperature, density, and stream velocity will vary. Pressure and temperature are related to combustion chamber values, P_c and T_c, by the condition for isentropic flow

$$S(P, T) = S(P_c, T_c)$$

The density at any point is

$$\rho = \frac{P}{RT}$$

which is a function of composition, since R is the gas constant (kJ/kg K) and involves the molecular weight, a function of composition. The stream velocity V is

$$V = \sqrt{2[h(P_c, T_c) - h(P, T)]}$$

written to emphasize the dependence of enthalpy on pressure as well as temperature. The primitive definition for the acoustic velocity in gases is

$$V_a = \sqrt{\left(\frac{\partial P}{\partial \rho}\right)_S} \qquad\qquad 14\text{-}5$$

As illustrated in Fig. 14.6, as the gas mixture moves from the combustion chamber along the nozzle, stream velocity increases, acoustic velocity decreases, and the curves cross at the throat. The computational difficulty that surrounds accurate identi-

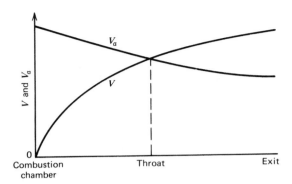

Figure 14.6 Variations in V and V_a in a rocket nozzle.

fication of throat conditions is precise calculation of the local acoustic velocity. We necessarily have to replace Eq. 14-5 with

$$V_a = \sqrt{\left(\frac{\Delta P}{\Delta \rho}\right)_S}$$ 14-6

The accuracy in V_a increases as $\Delta \rho$ becomes smaller and smaller, which in turn puts a high premium on accuracy in the calculations of mole numbers in the equilibrium mixture. One way to use Eq. 14-6 is to secure pressure and density values along the expansion, and at various points calculate V_a as the limit when $\Delta \rho$ approaches zero, as in calculus.

Figure 14.7 illustrates some results. The rocket is similar to the example in Fig. 14.5 but is now fueled with $C_{10}H_{22}$/liquid oxygen. This propellant combination produces much less in the way of altitude performance, as compared to the liquid H_2/liquid O_2 rocket. The two performance curves for shifting equilibrium with the mass flow rates calculated from Eqs. 14-3 and 14-4 show only small differences. This single example, of course, is no basis for any meaningful generalization. Figure 14.7 shows, as does Fig. 14.5, an enormous difference in ultimate rocket altitude between frozen flow and shifting equilibrium.

PROBLEMS

14.1   Calculate the adiabatic flame temperature for combustion of $C_{10}H_{22}$/air and $C_{10}H_{22}$/O_2(gas) initially at 1 atm and 300 K, with the products of combustion in chemical equilibrium. Using the programs sketched in Chapter 11, the products can be assumed to contain 6 or 10 species.

For constant volume combustion the results may be checked against Fig. 14.1. For constant pressure combustion the results may be checked against Fig. 14.2.

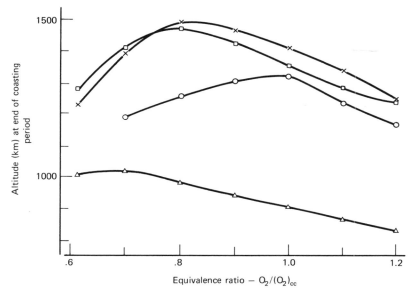

○————○ Complete combustion, flow rate from Eq. 14–3
△————△ Chemical equilibrium, frozen flow, flow rate from Eq. 14–3
×————× Chemical equilibrium, shifting equilibrium, from Eq. 14–3
□————□ Chemical equilibrium, shifting equilibrium, flow rate from Eq. 14–4

Figure 14.7 Single-stage $C_{10}H_{22}$/liquid O_2 vertical rocket. P_c = 20 atm; P_e = 1 atm; throat area = .056 m^2; M_o = 13 605 kg; M_p = 12 245 kg; π = .9.

This exercise provides practice in the essentials of the techniques for introducing chemical equilibrium in combustion reactions. In each case, the energy equation is dealt with in the form

$$U_r(T_r) = U_p(T)$$

where r and p denote reactants and products, and T_r is the temperature of the reactants. The method outlined in Section 2.7 must be employed.

14.2 Calculate the performance of an ideal Otto cycle engine (Fig. 5.1) assuming:
   (*a*) Complete combustion,
   (*b*) Equilibrium with 6 species in the products,
   (*c*) Equilibrium with 10 species in the products.
The results may be checked against Fig. 14.3.
   The program can then be refined further by including progressive combustion

and still further by a more realistic treatment of the exhaust/intake processes, which takes into account engine geometry and speed.

The extension to Diesel cycles follows with only minor changes.

14.3   In the case of rockets, the differences between computations based on complete combustion and on chemical equilibrium are more dramatic than for air-breathing engines, since rockets use oxygen to support combustion.

Note that various assumptions are available for treatment of the flow through the expansion nozzle. When the area at the throat cross section is given, the thrust developed depends directly on mass flow rate, and the latter may be calculated in a number of ways, all of which may be expected to give differing results. Similarly, thrust depends on exit section gas velocity, which may be computed in several ways, depending on the assumed behavior of the gases as they pass from the combustion chamber to the exit section.

For hydrocarbon/liquid O_2 mixtures, the methods for equilibrium calculations are given in Chapter 11, and for liquid H_2/liquid O_2 mixtures, the methods are given in Chapter 12.

15
SOLID FUELS AND POWDERS

15.1 INTRODUCTION

In this chapter the following aspects of burning in powders and solid fuels are examined: the adiabatic flame temperature, solid fuels as rocket propellants, and powders as propellants.

Powders and solid fuels in the form of powder castings may decompose in three ways. Combustion will occur when they are burned in the presence of air. Most powders burn quietly, with little flame or smoke. When burned in a confined space in the absence of oxygen, the burning is explosive. Decomposition takes place at the surface of the powder grains, the decomposing layer moving at some rate of travel into the body of the powder. A powder casting will behave similarly. The pressure in the vessel rises rapidly, causing an explosive effect. Detonation is faster than explosion, just as explosion is faster than combustion. Detonation is violent; the entire mass of powder decomposes almost instantaneously.

Explosion occurs in rocket motors and in gun barrels. Detonation is essential for blasting purposes. In guns, detonation will rupture the powder chamber.

Powders and fuel castings can produce very high pressures by virtue of their large density, as compared to gaseous reactants.

15.2 THE ADIABATIC FLAME TEMPERATURE

To describe the routine that can be employed to compute the temperature in the gaseous products following adiabatic combustion of a powder in a confined space, we shall examine what happens when cellulose decanitrate

$$C_{24}H_{30}O_{20}(NO_2)_{10}$$

is burned. Note first that 63 atoms of oxygen would be required to oxidize all the carbon and hydrogen to CO_2 and H_2O, but only 40 oxygen atoms are available. In view of this deficiency, the product gas mixture can be represented quite accurately by

$$n_1 CO + n_2 CO_2 + n_3 H_2O + n_4 H_2 + n_5 N_2 \qquad \text{15-1}$$

and chemical equilibrium is maintained by the so-called 'water gas' reaction

$$CO + H_2O \rightleftharpoons CO_2 + H_2 \qquad\qquad 15\text{-}2$$

For constant volume adiabatic combustion the energy equation for the flame temperature T is

$$U_{powder} = U_p(T) \qquad\qquad 15\text{-}3$$

Because of the large density of powders, assumed to be 1 g/cm^3 unless otherwise specified, as compared to air, the mass of air in the reactants may be ignored. The internal energy of the powder can be assigned a number consistent with the values for gases in Table E.2, using the thermodynamic definition for the standard heat of formation for cellulose decanitrate,

$$\Delta h_f^\circ = (h_{powder} - 24h_C - 15h_{H_2} - 20h_{O_2} - 5h_{N_2})_{298} \qquad\qquad 15\text{-}4$$

From Table E.4,

$$\Delta h_f^\circ = -2\ 803\ 000 \ \text{kJ/kmol}$$

Inserting the proper enthalpy values in Eq. 15-4

$$h_{powder}(298) = 11\ 226\ 000 \ \text{kJ/kmol}$$

For solids, $u \approx h$, because the Pv terms are small; hence the energy equation

$$11\ 226\ 000 = U_p(T) \qquad\qquad 15\text{-}5$$

can replace Eq. 15-3. For the mole numbers that occur in the internal energy of the product mixture, we have five equations

$$
\begin{aligned}
n_1 + 2n_2 + n_4 &= 40 \\
n_1 + n_2 &= 24 \\
n_3 + n_4 &= 15 \\
n_5 &= 5 \\
K = \frac{n_2 n_3}{n_1 n_4}
\end{aligned}
\qquad\qquad 15\text{-}6
$$

Table C.2 provides the equation with which to secure a value for K for an assumed T. Eliminating three of the four unknown mole numbers by substitution in the mass action equation, a simple quadratic equation results, and calculation of the equilibrium composition is simple. Note that the total mole number N

$$N = n_1 + n_2 + n_3 + n_4 + n_5 = 24 + 15 + 5 = 44$$

is constant, independent of the composition. Also, the composition is independent of pressure, since P does not occur in Eqs. 15-6. The total mole number is constant because there is no mole number change in Eq. 15-2. Solving Eq. 15-5 by the half-interval search scheme discussed in Appendix B, the flame temperature is 3225 K, and the mole fractions are

$$x_{CO} = 0.420$$
$$x_{CO_2} = 0.126$$
$$x_{H_2} = 0.103$$
$$x_{H_2O} = 0.238$$
$$x_{N_2} = \underline{0.114}$$
$$1.001$$

The corresponding total pressure can be estimated from the ideal gas equation

$$P = \frac{NRT}{V}$$

Suppose 1 mol of powder is burned, then $N = 44$ mol, the mass of powder is 1098 g and occupies a volume of 1098 cm^3. Suppose the powder occupied one-tenth of the container volume before ignition (this is referred to as a *loading* of 0.1), then the container volume V is

$$V = 10 \times 1098 = 10\,980 \text{ cm}^3$$

and $P = 1060$ atm, using

$$R = 82.057 \frac{\text{cm}^3 \text{ atm}}{\text{mol K}}$$

That pressure prediction is sufficiently high to warrant an alternative computation using the *co-volume* equation

$$P = \frac{RT}{v - b} \tag{15-7}$$

where v is the volume per mole

$$v = \frac{V}{N} = \frac{10\,980}{44} = 250 \text{ cm}^3/\text{mol mixture}$$

The b value can be approximated by weighting the b_i for each gas by its mole fraction,

$$b = \Sigma x_i b_i \qquad\qquad 15\text{-}8$$

Taking b_i values from Appendix A,

Gas	b_i cm^3/mol	x_i	$x_i b_i$
CO	31	0.420	13.02
CO_2	31.3	0.126	3.94
H_2	21.7	0.103	2.24
H_2O	18.7	0.238	4.45
N_2	30	0.114	3.42
			27.1 cm^3/mol mixture

and

$$P = 1060 \text{ (atm)} \frac{250}{250 - 27.1}$$

$$= 1190 \text{ atm}$$

Presumably, this second value is a better approximation than the earlier number. If the flame temperature calculation is repeated for a fully dissociated mixture containing

$$CO, CO_2, H_2, H_2O, O_2, N_2, NO, O, H, OH$$

the new temperature will be found to be within 1% of 3225 K. The reason for the agreement is the inhibiting effect of high pressure on dissociation reactions.

When the amount of oxygen in the powder molecule approaches, or exceeds, the amount required to oxidize completely all C and H to CO_2 and H_2O, the product gas mixture could be represented by the following species:

$$CO, CO_2, H_2, H_2O, O_2, N_2$$

The method described for cellulose decanitrate can be applied to any powder for which the heat of formation is known.

15.3 SOLID FUELS AS ROCKET PROPELLANTS

Powders, cast into a solid piece, are employed as rocket fuels. As with liquid fuels, decomposition produces a gas mixture at high pressure and temperature which develops a reactive thrust force as it flows through a convergent-divergent nozzle. There is a significant difference between liquid and solid fuels as used in rockets. In liquid fuel rocket motors, the rate of gas flow through the nozzle is controlled by the rates at which fuel and oxygen are pumped into the combustion chamber. (One of the major problems in the design of liquid fuel rocket motors is to match the rate of burning and the rates of pumping. This problem has not been discussed, for its solution carries the analysis into the field of reaction kinetics.) In a solid fuel rocket motor the rate of decomposition depends on the composition of the fuel, the geometry of the casting, and the prevailing pressure and temperature in the combustion chamber.

Figure 15.1 illustrates a solid fuel rocket motor, with the fuel cast into a cylinder, which burns like a cigarette when ignited over the entire exposed face. In the examination of the performance of the motor, we must consider two flow rates: $\dot{m}_g$, the mass rate of gas flow through the nozzle; and $\dot{m}_d$, the mass rate of decomposition of the fuel. From Eq. 7-28

$$\dot{m}_g = \Gamma A t P c \sqrt{\frac{MW}{RTc}} \qquad 15\text{-}9$$

where At is the nozzle throat cross-sectional area, MW is molecular weight of the gases, Tc is the combustion chamber temperature, Γ is a function of the gas specific heat ratio, (see Eq. 7-28), and Pc is the combustion chamber pressure. As we discovered in the previous section, Tc is not influenced by Pc for fuels with deficient oxygen. Consequently, if Tc is assumed to be virtually constant, then the mass flow rate in the nozzle portion of the motor is directly proportional to pressure. (Equation 15-9 applies to flow systems that reach the acoustic velocity at some cross section At, which is true of solid fuel motors.)

The mass rate of decomposition can be formulated as

$$\dot{m}_d = \rho A v_s \qquad 15\text{-}10$$

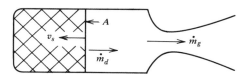

Figure 15.1 Solid fuel rocket.

where ρ is the casting density, A is the surface area over which decomposition is taking place, and v_s is the rate (cm/s, say) at which the decomposing zone moves into the fuel casting in the direction normal to A. The *surface regression rate,* v_s, is found by experiment to correlate with chamber pressure according to

$$v_s = aP_c^n \qquad\qquad 15\text{-}11$$

where a and n are approximately constant. Their magnitudes are fixed by the composition of the fuel and can therefore be adjusted by altering the ingredients, such as powder, binder, and additives, that go into the casting. Clearly, for steady rocket motor operation, we must have

$$\dot{m}_g = \dot{m}_d \qquad\qquad 15\text{-}12$$

Figure 15.2 illustrates the two possibilities that can arise. In Fig. 15.2*a*, the operation is unstable if $n > 1$ in Eq. 15-11. For if the pressure rises above the value associated with the intersection that is the desired operating point, the rate of decomposition will increase more rapidly than the rate of gas flow, and the motor will (theoretically) explode. On the other hand, in Fig. 15.2*a*, if the pressure drops below the intersection, the result will be a steady decrease in rate of decomposition, and the surface reaction will (theoretically) cease. When $n < 1$, as in Fig. 15.2*b*, the operation is stable.

The manufacture of solid powder fuel castings is a large subtechnology. Propellants can be produced with n values covering the range 0.2 to 0.7 and with surface regression rates which can vary widely; for most rockets, v_s falls in the range of 0.25 to 1.5 cm/s.

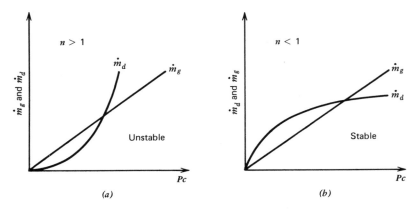

Figure 15.2 Solid fuel rocket, showing stable or unstable motor behavior, depending on the value of n in Eq. 15-13.

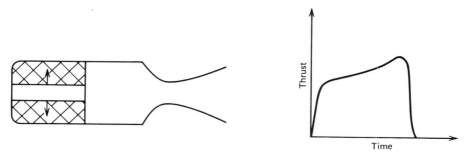

Figure 15.3 Solid fuel rocket motor and associated thrust.

Solid fuels have some distinct advantages when compared with cryogenic fuels such as liquid oxygen and hydrogen. Easier long-term storage is an obvious advantage. Whereas liquid fuels are pumped at steady rates, and therefore produce constant thrust, powders may be cast into a variety of shapes, so that the surface area of the decomposing fuel, A in Eq. 15-12, is not necessarily constant but may change with time in some predetermined fashion. Figure 15.3 is an example; when the fuel casting is a hollow cylinder, and the decomposing surface extends over the inner surface as indicated, A will increase with time, as will thrust. A variety of thrust-time profiles are available. However, to achieve them, the intended surface area must be ignited promptly and reliably, an assignment not easily guaranteed.

15.4 POWDERS AS PROPELLANTS

Powders have been used for centuries to propel missiles from cannons and guns. That long history has not eliminated the need for empirical information about the decomposing characteristics of a powder when accurate predictions about the flight path of the missile is the objective. *Interior ballistics* is concerned with the motion of a missile as it moves in a gun barrel; we shall examine the problem here, with simplifications.

A missile moves under the action of the pressures on its surfaces and the forces acting between missile and barrel. Consider first the pressure that the gases produce by decomposition of the powder exert. In Fig. 15.4a, a gas particle with mass m and speed c is moving toward a stationary piston, surface area A, a distance L from the closed end of the cylinder. If the collision with the piston is perfectly elastic, the momentum change of the particle will be

$$2mc$$

If there are N particles in the cylinder, then we may assume that one-third are moving normal to the piston; that is, there is no preferred direction, a circumstance supported

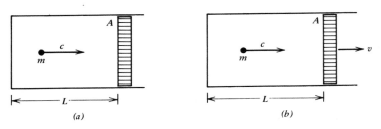

Figure 15.4 (a) Stationary piston. (b) Moving piston.

by the observation that the pressure on the inner surface of the cylinder and piston is everywhere uniform. The number of particles colliding with the stationary piston per unit time is

$$\frac{N}{3}\frac{c}{2L}$$

so that the momentum change in the gas per unit time is

$$\frac{1}{3}Nmc^2\frac{1}{L} \tag{15-13}$$

This is also the force the gas exerts on the piston, and dividing by the area A, we have at once the pressure exerted by the gas on the piston,

$$P = \frac{1}{3}\frac{Nmc^2}{V} \tag{15-14}$$

where $V = AL$ is the cylinder volume. Equation 15-14 is one of the early triumphs of kinetic theory. Since $PV = NRT$; the speed of the particles, can be found from

$$c = \sqrt{3\frac{RT}{MW}}$$

where R, the universal gas constant, $= 8.314 \text{ kJ/kmol K}$, and MW is the molecular weight of the gas. Thus

$$c = 158\sqrt{\frac{T}{MW}} \qquad \text{(m/s)} \qquad (T \text{ in degree K}) \tag{15-15}$$

At 3000 K, for example, the speeds are

Gas	MW	c(m/s)
H_2	2	6000
H_2O	18	2000
CO	28	1600
CO_2	44	1300

In a mixture of gases, all the particles are at the same temperature, so that Eq. 15-15 applies for each gas.

Now consider Fig. 15.4b, a different situation, where the piston is moving to the right with speed v. The analysis just given must be revised. First, examine the momentum change resulting from the collision of a particle with mass m moving with speed c normal to the piston, measured with respect to the cylinder wall. To an observer moving with the piston, that particle appears to approach with speed $c - v$ and to rebound with the same speed, since the collision is elastic. The observer records for each collision the momentum change

$$2m(c - v) \qquad\qquad 15\text{-}16$$

To a stationary observer, the particle appears to approach the piston with speed c, and since both observers must report that same momentum change, in the stationary frame of reference the particle rebounds with speed $c - 2v$. Since the piston is moving, there is work transfer between the gas and the piston, and a change in the energy of the gas. The energy in the gas is entirely kinetic. As before, the number of particles that collide with the moving piston in time interval dt is

$$\frac{N}{3}\frac{c}{2L}dt$$

and in that time interval

$$\text{Decrease in kinetic energy} = \frac{N}{3}\frac{c}{2L}\frac{m}{2}[c^2 - (c - 2v)^2]dt$$

$$= \frac{1}{3}Nmc^2\frac{v}{L}\left(1 - \frac{v}{c}\right) \qquad\qquad 15\text{-}17$$

The work done by the gas on the piston, W, during the time interval dt is

$$W = pAv\,dt \qquad\qquad 15\text{-}18$$

where p is the pressure the gas exerts on the piston. Equating the decrease in kinetic energy with the work transfer, we have, making use of Eq. 15-14,

$$p = P \left(1 - \frac{v}{c}\right) \qquad\qquad 15\text{-}19$$

Here P is the pressure that would be recorded on any stationary surface, the *static* pressure. (When the piston moves inward, the sign in Eq. 15-19 changes from minus to plus.) Since muzzle velocities in guns and cannons can be of the order of 1000 m/s, the correction factor v/c can be substantial and must enter into a careful analysis of the motion of the missile in the gun barrel.

As an illustrative example, suppose the following conditions are fixed:

Mass of the bullet = 45.3 g
Mass of powder = 12.9 g
Length of gun barrel = 0.762 m
Cross-sectional area of barrel = 3.22 cm^2

The powder is cellulose decanitrate, as in Section 15.2. In real guns, the barrel contains internal helical grooving that imparts spin and stabilizes the attitude of the projectile during free flight after it leaves the barrel. A force, F, is required to start the bullet moving, and some lesser force, F', is needed to overcome friction between bullet and barrel. These items will be ignored. To complete a statement of the problem, we need to specify the loading factor, Δ,

$$\Delta = \text{loading factor}$$

$$= \text{initial powder volume/powder chamber volume}$$

and we must specify in some way the burning characteristic of the powder. The loading factor may be chosen anywhere between zero and unity. Since the powder mass is fixed, so is its volume, so that Δ fixes the powder chamber volume. At any moment during the motion, the static pressure behind the bullet is calculated from

$$P = \frac{RT}{V/N - b} \qquad\qquad 15\text{-}20$$

in which V = volume occupied by powder gases

= powder chamber volume + barrel volume behind bullet − volume of unburned powder,

N = moles of gas,

b = co-volume, Eq. 15-8.

As for the burning characteristics of the powder, the simplest arrangement is to assume $n = 0$, in Eq. 15-11, with a in that equation constant. The burning rate is then constant. However, we can imagine that the burning rate is capable of manipulation. Clearly, a finely divided powder with large surface area/volume ratio can be expected to burn more rapidly than can the same powder with smaller ratio.

Figure 15.5 illustrates the variations with time for the static pressure, P; the pressure ratio, p/P, from Eq. 15-19; the velocity of the bullet, v, for $\Delta = 0.6$; and a total burning time of 1/1000 second. The muzzle velocity is about 800 m/s. The sharp rise in P with time is characteristic of interior ballistic calculations. The form of the curve will be changed if the burning rate is a function of P.

The computation proceeds by dividing the burning time into a number of intervals of equal size. (The effect of interval size on results needs to be explored.)

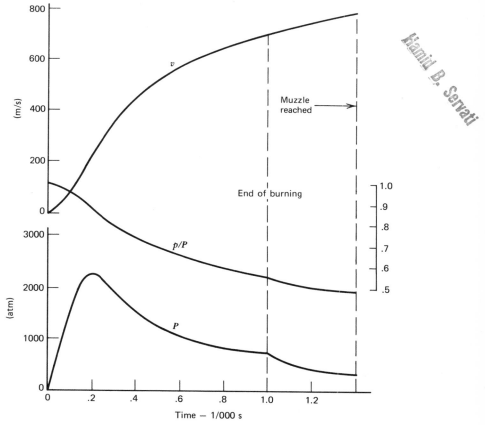

Figure 15.5 Gun firing calculations; $\Delta = 6$; burning time = 1/1000 sec.

During the burning period, the static pressure P will change as the result of burning and bullet movement. The computations bear some resemblance to progressive burning, discussed in Section 5.9, except that in an engine, piston position is a known function of time, whereas in a gun, the object is to determine bullet position as time passes. With the bullet momentarily stationary, the energy equation is

$$U_p(T) - U_p(T + \Delta T) = Q \qquad\qquad 15\text{-}21$$

where U_p denotes internal energy of the combustion gases, and Q is the energy released

$$Q = fM_p u_{\text{powder}} \qquad\qquad 15\text{-}22$$

with f denoting the fraction of initial powder mass M_p, which burns in time interval Δt, producing temperature change ΔT. When the bullet moves, the energy equation is

$$U_p(T) - U_p(T + \Delta T) = pAv\ \Delta t \qquad\qquad 15\text{-}23$$

with p calculated from Eq. 15-19, v denoting bullet velocity, and A denoting barrel cross-sectional area. Note that in Eq. 15-21 the energy terms are associated with different masses of gas, while in Eq. 15-23 the mass of gas is identical for both terms. For a horizontal barrel, the change in bullet velocity is derived from

$$M_b\ \Delta v = (p - P_{\text{atm}})A\ \Delta t \qquad\qquad 15\text{-}24$$

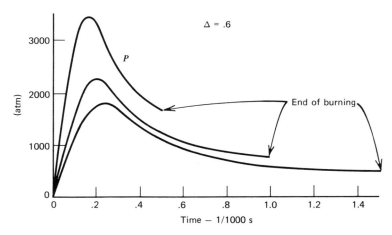

Figure 15.6 Effect of burning time on statical pressure P.

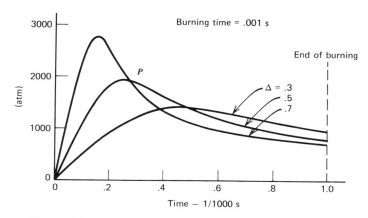

Figure 15.7 Effect of loading factor on statical pressure *P*.

with P_{atm} denoting ambient pressure, and M_b denoting the bullet mass. Note that the gas does not expand isentropically as the bullet moves. This is a consequence of Eq. 15-19. When a gas is compressed and then expanded to its original volume by a piston moving at finite velocity, the work required for compression is not regained during expansion, and the gas is warmer at the end of the expansion stroke. The gas does not perform a cycle.

The loading factor can be altered, for a fixed powder mass, by changing the powder chamber volume. The burning time may be altered for the same powder by changing the size of powder grains. The separate effects of these parameters on the variation of static pressure with time are illustrated in Figs. 15.6 and 15.7. The curves are drawn only for the burning period.

Including the co-volume factor *b* increases the computed static pressure, as in Eq. 15-20, while the inclusion of the bullet velocity decreases the pressure *p* acting on the piston, as in Eq. 15-19. The effects of the latter are greater than those of the former, as the numbers in Table 15.1 indicate. The muzzle velocities in the first

TABLE 15.1
LOADING FACTOR = 0.6
COMPARISON OF MUZZLE VELOCITY (m/s) CALCULATIONS

Burning Time (s)	Ignoring Co-volume and Bullet Speed	Including Co-volume and Bullet Speed
0.0005	1030	856
0.00075	977	822
0.001	929	789
0.00125	884	758
0.0015	842	731

TABLE 15.2

Burning Time (s)	Loading (Δ)				
	0.3	0.4	0.5	0.6	0.7
0.0005	813*	833	846	854	857
	(2204)†	(2574)	(2981)	(3483)	(4133)
0.00075	791	806	815	818	821
	(1684)	(1974)	(2303)	(2696)	(3224)
0.001	771	782	788	789	787
	(1403)	(1649)	(1925)	(2282)	(2774)
0.00125	746	757	760	759	756
	(1213)	(1427)	(1665)	(1960)	(2396)
0.0015	724	732	732	730	726
	(1078)	(1282)	(1513)	(1807)	(2160)

*Muzzle velocity in meters per second.
†Peak pressure in atmospheres.

column are based on static pressure, calculated from the ideal gas equation $PV = NRT$, acting to move the bullet. In the second column, the co-volume factor and the bullet speed correction are both included.

Table 15.2 illustrates how loading and burning time combine to produce various muzzle velocities and peak pressures. As the burning time increases, both muzzle velocity and peak pressure decrease for all loading factors. On the other hand, for a given burning time, the muzzle velocity exhibits a maximum, while the peak pressure increases steadily with increasing loading factor. Short burning time and large loading will produce the highest muzzle velocity, but the price paid is high peak pressure, with the possibility of damage to the barrel and powder chamber.

PROBLEMS

15.1 Verify the flame temperature of 3225 K for the combustion of $C_{24}H_{30}O_{20}(NO_2)_{10}$, and show why it is independent of loading.

15.2 Real gases can absorb energy in three forms: translation, rotation, and vibration. Is the correction factor developed in Eq. 15-19 valid for real gases?

APPENDICES

APPENDIX A THE SI UNIT SYSTEM AND THERMODYNAMIC PROPERTIES OF GASES AND MIXTURES OF GASES

In the SI unit system (Système International d'Unités), the base units are arbitrarily chosen:

Length, meter (m)
Time, second (s)
Mass, kilogram (kg)
Temperature, Kelvin (K)
Amount of substance, mole (mol)

The derived units, then, are as follows.

The newton (N) is the force that accelerates a mass of one kilogram at the rate of one meter per second per second.

The joule (J) is the work done when the point of application of one newton is displaced a distance of one meter in the direction of the force. (A joule is a relatively small quantity of energy. In terms of commonplace experience, one kilojoule is the energy released by burning one "strike-anywhere" kitchen match.)

The watt (W) is the power that gives rise to the production of energy at the rate of one joule per second.

The pascal (Pa) is the pressure exerted by one newton distributed over one square meter.

In the SI unit system, Newton's second law does not require a dimensional conversion factor. The following conversion factors may be helpful:

1 Btu = 1050 J
1 cal = 4.18 J
1 in. = 2.54 cm
1 ft = 0.305 m
1 mile = 1.61 km
1 in.^{2} = 6.45 cm^2
1 ft^2 = 0.0929 m^2
1 in.^{3} = 16.4 cm^3 = 0.0164 liter
1 ft^3 = 0.0283 m^3
1 lbf/in.^{2} = 6890 Pa
1 atm = 101 000 Pa
1 lbm = 0.454 kg
1 lbf = 4.45 N
1 degree R = 5/9 degree K
1 hp = 0.746 kW
1 hp/in.^{3} = 45.5 kW/liter
32.2 ft/s^2 = 9.80 m/s^2

In the ideal gas equation, $Pv = RT$

$$R = 8.314 \text{ kJ kmol}^{-1} \text{ K}^{-1}$$

The following are more convenient R values when dealing with engines:

$$R = 82.06 \frac{\text{cm}^3 \text{ atm}}{\text{mol K}}$$

$$= 0.08206 \frac{\text{liter atm}}{\text{mol K}}$$

The mole is the amount of substance of a system that contains as many elementary entities as there are carbon atoms in 0.012 kg or 12 g of carbon 12. The elementary entities must be specified and may be atoms, molecules, ions, electrons, other particles, or specified groups of such particles.

In the SI unit system, the mole replaces the gram mole (gmol).

The number of entities in a mole is 6.022×10^{23}, which is also known as Avogadro's constant.

1 mol of O_2 has a mass of 32 g
1 mol of H_2 has a mass of 2 g
1 mol of N_2 has a mass of 28 g, etc.
1 mol of 0 has a mass of 16 g, etc.

The first three are known as molecular weights; the last is known as an atomic weight.

The internal energy, u, and enthalpy, h, are related:

$$h = u + Pv \qquad \text{A-1}$$

For ideal gases, we can also write

$$h = u + RT \qquad \text{A-2}$$

Internal energy and enthalpy are defined by

$$du = c_v \, dT$$

$$dh = c_p \, dT \qquad \text{A-3}$$

and

$$c_p - c_v = R \qquad \text{A-4}$$

The ratio of the specific heats is designated by k

$$k = \frac{c_p}{c_v} \qquad \text{A-5}$$

The heat capacities c_p and c_v are weak functions of pressure but strong functions of temperature. Values for c_p° are tabulated in Table E.1,. The superscript $^\circ$ means that the values are for 1 atm pressure.

Internal energy and enthalpy are strong functions of temperature:

$$u(T) = \int_{T_o}^{T} c_v \, dT + u(T_o) \qquad\qquad\qquad \text{A-6}$$

$$h(T) = \int_{T_o}^{T} c_p \, dt + u(T_o) + R T_o \qquad\qquad \text{A-7}$$

The datum temperature T_o and $u(T_o)$ or $h(T_o)$ may be assigned arbitrarily. A convenience is achieved when the datum values are established in a systematic fashion, as described in Appendix E. Enthalpy values for gases are listed in Table E.2,. Internal energy values can be secured using Eq. A-2.

Entropy is a more complicated function and is derived from the first law,

$$ds = \frac{du + P \, dv}{T}$$

which can take three forms when $Pv = RT$ is introduced,

$$\left.\begin{aligned}
ds &= c_v \frac{dT}{T} + R \frac{dv}{v} \\[4pt]
ds &= c_p \frac{dT}{T} - R \frac{dP}{P} \\[4pt]
ds &= c_p \frac{dv}{d} + c_v \frac{dP}{P}
\end{aligned}\right\} \qquad\qquad \text{A-8}$$

Introducing the function $\phi(T)$, defined by

$$\phi(T) = \int_{T_o}^{T} c_p \frac{dT}{T} + \phi(T_o) \qquad\qquad \text{A-9}$$

the entropy difference between two states 1 and 2 can be written as

$$s_2 - s_1 = \phi(T_2) - \phi(T_1) - R \ln \frac{P_2}{P_1} \qquad\qquad \text{A-10}$$

Tabulated values of $\phi(T)$ are listed in Table E.3. Since k, Eq. A-5, is not a strong function of temperature, Eqs. A-8 may be integrated, treating k as a constant:

$$Pv^k = \text{constant}$$
$$TP^{(k-1)/k} = \text{constant} \ (s = \text{constant}) \qquad\qquad \text{A-11}$$
$$Tv^{k-1} = \text{constant}$$

Equations A-11 will give quite accurate values over large temperature ranges. Equation A-10 is, of course, more precise, since the $\phi(T)$ function takes into account the temperature variation of c_p.

In a mixture of gases, with n_i denoting the mole number of gas i, the total mole number N is

$$N = \Sigma n_i$$

The internal energy, U, enthalpy, H, and entropy, S, of a mixture are obtained by adding the contributions from each constituent gas,

$$U(T) = \Sigma n_i u_i(T)$$
$$H(T) = \Sigma n_i h_i(T) \qquad\qquad \text{A-12}$$
$$S(P, T) = \Sigma n_i s_i(p_i, T)$$

where the partial pressure, p_i, is

$$p_i = \frac{n_i}{N} P \qquad \text{and} \qquad P = \Sigma p_i \qquad\qquad \text{A-12}$$

The internal energy, enthalpy, and entropy of a mixture, per mole of mixture, can be expressed by introducing the mole fraction, x_i

$$x_i = \frac{n_i}{N}, \qquad \Sigma x_i = 1 \qquad\qquad \text{A-13}$$

and then

$$u(T) = \Sigma x_i u_i(T)$$
$$h(T) = \Sigma x_i h_i(T) \qquad\qquad \text{A-14}$$
$$s(P, T) = \Sigma x_i s_i(p_i, T)$$

When the heat capacity of a mixture of gases is written with capital C's,

$$C_p = \Sigma n_i c_{p\,i}, \qquad C_v = \Sigma n_i c_{v\,i} \qquad\qquad \text{A-15}$$

the proper units are J/K or kJ/K. When written with lowercase $c's$

$$c_p = \Sigma x_i c_{p\,i}, \qquad c_v = \Sigma x_i c_{v\,i} \qquad\qquad \text{A-16}$$

the proper units are J mol^{-1} of mixture K^{-1} or kJ mol^{-1} of mixture K^{-1}. Equations A-15 and A-16 apply only to mixtures with fixed composition. If the composition can change, as the result of chemical reaction, then the heat capacities are identified by C_p^* and C_v^* and are formulated from their basic definitions:

$$C_p^* = \frac{d}{dT} H(T) = \frac{d}{dT} \Sigma n_i h_i = \Sigma n_i \frac{dh_i}{dT} + \Sigma h_i \frac{dn_i}{dT}$$

$$= C_p + \Sigma h_i \frac{dn_i}{dT} \;, P \text{ constant} \qquad\qquad \text{A-17}$$

and

$$C_v^* = C_v + \Sigma u_i \frac{dn_i}{dT} \;, V \text{ constant} \qquad\qquad \text{A-18}$$

Under certain circumstances the second terms in Eqs. A-17 and A-18 may be substantially larger than C_p and C_v. This is discussed in Chapter 13.

The molecular weight of a mixture, MW, is simply related to the molecular weights of the constituent gases,

$$MW = \Sigma x_i MW_i \qquad \text{A-19}$$

As a general rule, when the composition of a mixture of ideal gases is constant, any property of the mixture per mole of mixture (when appropriate) is derived by weighting the corresponding property of each constituent according to each mole fraction.

In the region of small specific volume, the ideal gas equation is modified and appears as

$$P(v - b) = RT \qquad \text{A-20}$$

where R is still the universal constant, and b is the *co-volume* which takes into account the volume occupied by the gas particles themselves. Values for b are shown to be one-third the critical specific volume and hence vary among gases, as the following values show:

Gas	$b(\text{cm}^3/\text{mol})$
Air	30.8
CO	31.0
H_2	21.7
N_2	30.0
O_2	24.7
CO_2	31.3
H_2O	18.7

For a mixture of gases, the b value can be approximated by

$$b_{\text{mixture}} = \Sigma x_i b_i \qquad \text{A-21}$$

The approximate character of Eq. A-21 can be observed by noting that air has a larger b value than do both O_2 and N_2.

Air consists of nitrogen, oxygen, and small amounts of several other gases. For engineering calculations, the latter may be ignored and mole fractions assigned to nitrogen and oxygen as follows:

$$x_{N_2} = 0.79, \qquad x_{O_2} = 0.21$$

The ratio $x_{N_2}/x_{O_2} = 3.76$, so that each oxygen molecule is accompanied by 3.76 nitrogen molecules.

In dealing with the flow of gases, it is useful to introduce the concept of *stagnation temperature* and to distinguish it from the *stream temperature*. The former is measured with a thermometer that is stationary with respect to the moving gas, while the latter would be measured with an infinitesimally small thermometer that moves with the gas. Stagnation temperatures are indicated with a superscript, T°, while

T denotes stream temperature. For a gas moving with velocity V, the relationship between the two temperatures is given by the energy equation

$$h(T) + \tfrac{1}{2}V^2 = h(T^\circ)$$

since the kinetic energy is converted into thermal energy as the fluid decelerates to zero velocity as it impacts against the stationary thermometer. The above equation contains enthalpy terms per unit mass. Writing them as $c_p T$ products, we get

$$T^0 = T + \frac{V^2}{2c_p}$$

As an example, for air, c_p = 29.2 J/mol K at room temperature, or about 1 J/g K, since the molecular weight is 28.96. Thus

$$T^0 = T + \frac{V^2}{2000} \text{ (K)}, \qquad \text{with } V \text{ in } (\text{m/s})^2$$

Consequently, at a velocity of 45 m/s, the difference between T and T° amounts to 1 °C. When V becomes large, the variation of c_p with T must be taken into account to reach an accurate value for the temperature difference.

APPENDIX B   SOLUTIONS FOR $F(T) = 0$

Solutions for $F(T) = 0$ can be secured in a variety of ways. Two simple and well-known methods are described here.

In order to produce a successful algorithm, *something* must be known about the form of $F(T)$. The more we know, the easier it will be to write the algorithm and the better the chance that it will be successful in all situations. When $F(T) = 0$ arises in thermodynamic analysis, and T represents absolute temperature, we know at once that we are concerned only with positive T values and that $F(T)$ must be single valued. The forms for $F(T)$ encountered frequently, and prominently, are energy equations, such as

$$F(T) = H_p(T) - H_r(T_r)$$

or

B-1

$$F(T) = U_p(T) - U_r(T_r)$$

or isentropic changes of state from 1 to 2, for which

$$F(T) = S(P_2, T) - S(P_1, T_1)$$

or

B-2

$$F(T) = S(V_2, T) - S(V_1, T_1)$$

In both cases, T is unambiguous; there is only one solution, so that in the range of interest, T positive, $F(T)$ is a single valued function. We know more. In the case of Eqs. B-1, we know T is greater than T_r, when T_r represents the initial temperature of the reactants. When Eqs. B-2 describe an expansion, then we know T must be less than T_1. Information of this sort simplifies construction of the program.

B.1 The Half-Interval Search

Before the search by halving can be started, we must bracket the solution, as in Fig. B.1. Because $F(T)$ is single valued, when

$$F(T1) \times F(T2) < 0 \qquad \text{B-3}$$

the solution must lie between $T1$ and $T2$. The first task is to find $T1$ and $T2$. Once they are in hand we calculate $T3$

$$T3 = (T1 + T2)/2 \qquad \text{B-4}$$

and by examining either product

$$F(T1) \times F(T3) \text{ or } F(T2) \times F(T3) \qquad \text{B-5}$$

determine whether the solution lies in the interval between $T1$ and $T3$ or between $T2$ and $T3$. (There is always the remote possibility that $F(T3) = 0$, and the program can be arranged to deal with it.) The routine of halving the interval may continue as long as we choose.

When we know more about the behavior of $F(T)$, the program simplifies. For example, if we know whether the function increases or decreases with T we need only look at the sign of $F(T3)$.

Figure B.2 contains a flow chart for the half-interval search in which we know

$$F(T) \text{ increases with } T$$

$$F(TO) \text{ is negative.}$$

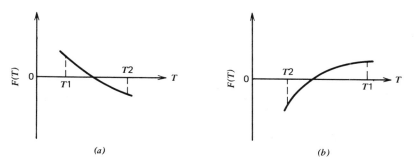

(a)   (b)

Figure B.1 Before the half-interval search can be started, the solution must be bracketed between $T1$ and $T2$.

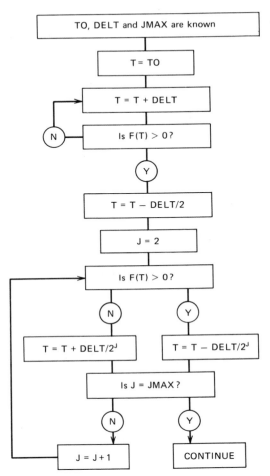

Figure B.2 Flow chart for half-interval search. We know: $F(T)$ increases with T, and $F(TO)$ is negative.

The program sets $T = TO$, and proceeds to increase T by $DELT$ until $F(T)$ becomes positive, at which point T is reduced to the midpoint of the interval which brackets the solution, and the half-interval search takes over. It ceases when the exponent J reaches $JMAX$. The latter is determined by the accuracy needed. If we wish to know T correct to within

$$\pm \text{ERROR}$$

then

$$\text{ERROR} = \text{DELT}/2^{\text{JMAX}}$$

or

$$JMAX = \log(DELT/ERROR)/\log(2) \qquad \text{B-6}$$

This leaves the magnitude of *DELT* to be decided. There is no definitive answer for the choice of *DELT*. By examining some hypothetical situations, the reader should quickly see that large *DELT* is preferred.

The alterations to the flow chart in Fig. B.2 if $F(T)$ decreases with T and/or $F(TO)$ is positive need no explanation.

B.2 Newton—Raphson Iteration

The strategy of the iteration is illustrated in Fig. B.3. Construct the tangent to the curve at $T1$. The tangent intersects the T-axis at $T2$, and by definition the slope at $T1$ is

$$F'(T1) = \frac{F(T1)}{T1 - T2}$$

which can be solved for $T2$

$$T2 = T1 - F(T1)/F'(T1) \qquad \text{B-7}$$

which is known as a recursion equation; $T2$, a better approximation to the solution than $T1$, can be computed from $T1$. By repeated application of Eq. B-7, the solution can be approached as closely as we desire.

As a general rule, the Newton-Raphson iteration converges more rapidly than does the half-interval search. But we need to know more about the character of $F(T)$ in order to use the Newton-Raphson iteration with assured success. For one thing, $F(T)$ must be differentiable. This is seldom a problem. Figure B.4 illustrates the sorts of problems which may arise, however. When, as in Fig. B.4*b*, the curve is concave upward, the solution can be found for any starting T value. Note that the solution is approached from above; successive approximations are each smaller than the preceding value. Trouble lurks in Fig. B.4*a*, where the curvature is concave downward. If we start at B, for example, the slope at that point may be such as to cause the tangent to

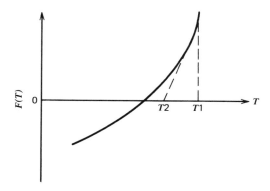

Figure B.3 Newton-Raphson iteration.

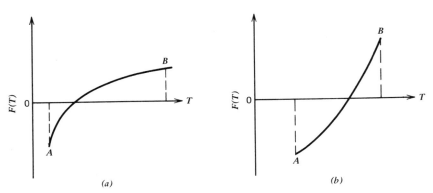

Figure B.4 Newton-Raphson iteration. In (*a*), starting the program at *B* may produce a negative *T*. This is avoided by starting at *A*. In (*b*) the program may be started at *A* or *B*.

intersect somewhere on the negative *T*-axis. If $F(T)$ contains $\log(T)$, as is the case for u, h, and ϕ functions (see Appendix C), the program will abort. This can be avoided by starting at *A*, and the solution is approached from below. Since the thermodynamic functions are computed with expressions that are not valid below 400 K or above 6000 K, *T* values outside those limits must be avoided. (Actually, the high temperature limit is seldom a real problem.) Figure B.5 shows a flow chart for the Newton-Raphson

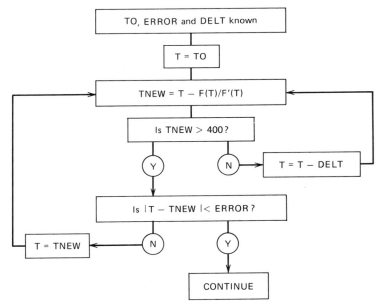

Figure B.5 Flow chart for Newton-Raphson iteration *TO* must be greater than 400 K.

iteration. The flow chart is arranged to carry the calculations to a solution for any TO, provided it is greater than 400 K, by systematically reducing T by DELT until the recursion expression produces an acceptable value.

But there are circumstances under which the flow chart in Fig. B.5 will not converge. This may occur if the derivative $F'(T)$ does not increase or decrease uniformly and continuously with T. When this condition is not met, $F(T)$, while monotonic, may be S-shaped. In this eventuality, some manipulations are required to discover an initial T value that will guarantee convergence. This topic is discussed in reference to Figs. 13.1 and 13.2, Section 13.2, which deals with the problems surrounding the application of the Newton-Raphson iterative scheme to mixtures of gases in a state of chemical equilibrium.

APPENDIX C   ALGEBRAIC   EXPRESSIONS   FOR   THERMODYNAMIC FUNCTIONS

For hand calculations, tabulated values of c_p, the constant pressure heat capacity; h, enthalpy; ϕ, the temperature portion of entropy; and K_p, the reaction constant, are adequate. For computer programs, algebraic expressions are required. This can be accomplished in several ways.

The functions are related:

$$\left.\begin{aligned} h(T) &= \int_{T_o}^{T} c_p \, dT + h(T_0) \\ \phi(T) &= \int_{T_o}^{T} \frac{c_p}{T} \, dT + \phi(T_0) \end{aligned}\right\} \qquad \text{C-1}$$

and for the reaction

$$\nu_1 A_1 + \nu_2 A_2 \rightleftharpoons \nu_3 A_3 + \nu_4 A_4$$

van't Hoff's equation (Eq. 10-14) provides the value for K

$$\begin{aligned} \ln K_p(T) &= \int_{T_0}^{T} \frac{\Delta H}{RT^2} \, dT + K(T_0) \\ &= \int_{T_0}^{T} \frac{\nu_3 h_3 + \nu_4 h_4 - \nu_1 h_1 - \nu_2 h_2}{RT^2} \end{aligned} \qquad \text{C-2}$$

With experimentally determined values of c_p, formerly developed from calorimetric data and more recently from analysis of spectroscopic measurements, tabulated values of the enthalpy and entropy function can be prepared.

For the purpose of this book, the following expression was chosen to represent the enthalpy

$$h(T) = A + BT + C \ln(T) \qquad \text{(kJ/kmol)}$$

and the remaining thermodynamic functions are then

$$c_p(T) = B + \frac{C}{T} \qquad (\text{kJ/kmol K})$$

$$c_v(T) = B - 8.314 + \frac{C}{T} \qquad (\text{kJ/kmol K})$$

$$u(T) = A + (B - 8.314)T + C \ln(T) \qquad (\text{kJ/kmol}) \qquad\qquad \text{C-3}$$

$$\phi(T) = B \ln(T) - \frac{C}{T} + D \qquad (\text{kJ/kmol K})$$

$$K_p(T) = \exp\left[\frac{a}{T} + \left(b + \frac{c}{T}\right)\ln(T) + d\right] \qquad (P \text{ in atm})$$

Three-constant expressions will not adequately fit the enthalpy and entropy functions, over the temperature range from 400 to 6000 K. Therefore, the range was subdivided and the coefficients evaluated as follows:

> From 1600 to 6000 K: A, B, and C were evaluated at 2500, 4000, and 5500 K from the enthalpy tables; D was evaluated at 3500 K from the entropy function; and d was evaluated at 3500 K from the reaction constant tables.

> From 400 to 1600 K: A, B and C were evaluated at 500, 1100 and 1600 K, and D was evaluated at 1600 K. (The reaction constants are not needed below 1600 K.)

Table C.1 lists the values of A, B, C, and D for 11 gases for the two temperature ranges. Table C.2 lists the values of a, b, c, and d for nine reactions.

When the temperature range is subdivided, it is important to have a precise "fit" at the point of subdivision. If this provision is not made, it is possible for an iterative routine to "stagger" about that point and not converge. The high temperature range coefficients were first determined, and the low temperature coefficients were adjusted to give enthalpy and entropy function values that coincide at 1600 K with the high temperature coefficient values.

A single equation for each thermodynamic function could be developed by starting with an enthalpy expression containing, say, six or seven constants. As a general rule, the smaller the temperature span, the better is the agreement between calculated and tabulated values. The important point to appreciate is that *any* algebraic expression, no matter how absurd, can be forced to pass through *any* n points of *any* curve, provided the expression contains n adjustable parameters. But if the chosen expression is indeed absurd, it will yield wildly inaccurate values in some or all of the intervals between the n points.

If the enthalpy expression is chosen as the starting point for the development of equations to compute thermodynamic functions, then

1. The enthalpy expression must give satisfactory agreement with tabulated values of $h(T)$ over the entire selected temperature range;
2. The expresssion for enthalpy must be differentiable to provide an equation for c_p;
3. $(c_p/T)\,dT$ must be integrable, in order to derive the expressions for $\phi(T)$; and
4. $(h/T^2)\,dT$ must be integrable to obtain the reaction constant equation.

TABLE C.1
$A, B, C,$ AND D COEFFICIENTS FOR EQS. C-3

Gas	A	B	C	D
		$400 \leqslant T \leqslant 1600$ K		
CO	299180.	37.85	−4571.9	−31.10
CO_2	56835.	66.27	−11634.0	−200.0
H	357070.	20.79	−7.9	−3.9
H_2	326490.	40.35	−8085.2	−121.0
H_2O	88923.	49.36	−7940.8	−117.0
N_2	31317.	37.46	−4559.3	−34.82
O	265120.	24.60	−2729.2	13.86
O_2	43388.	42.27	−6635.4	−55.15
OH	217810.	37.36	−5561.4	−44.06
NO	111050.	37.81	−2874.8	−15.70
N	326040.	17.19	5371.4	64.67
		$1600 \leqslant T \leqslant 6000$ K		
CO	309070.	39.29	−6201.9	−42.77
CO_2	93048.	68.58	−16979.0	−220.4
H	357010.	20.79	0	−3.82
H_2	461750.	46.23	−27649.0	−176.6
H_2O	154670.	60.43	−19212.0	−204.6
N_2	44639.	39.32	−6753.4	−50.24
O	298360.	23.17	−6910.3	21.81
O_2	127010.	46.25	−18798.0	−92.15
OH	298750.	42.86	−17695.0	−92.24
NO	138670.	39.92	−7061.8	−33.90
N	486400.	26.91	−18159.0	−20.31

TABLE C.2
COEFFICIENTS FOR CALCULATION OF REACTION CONSTANTS K_p WITH THE EQUATION

$$K_p = \left[\exp \frac{a}{T} + \left(b + \frac{c}{T} \right) \ln(T) + d \right]$$

VALID FOR $1600 < T < 6000$ K. *Pressures must be in units of atmospheres.*

Reaction	a	b	c	d
$H_2 + \frac{1}{2}O_2 \rightleftharpoons H_2O$	42450.	−1.0740	−2147.0	3.2515
$CO + \frac{1}{2}O_2 \rightleftharpoons CO_2$	33805.	0.7422	165.8	−16.5739
$2H \rightleftharpoons H_2$	33587.	0.5604	3327.0	−20.8683
$2O \rightleftharpoons O_2$	57126.	−0.0100	599.0	−16.3201
$2H + O \rightleftharpoons H_2O$	104702.	−0.5181	1480.0	−25.8073
$O + H \rightleftharpoons OH$	44216.	−0.1319	1298.0	−13.1303
$CO + H_2O \rightleftharpoons CO_2 + H_2$	−8645.	1.8162	2312.0	−19.8254
$O_2 + N_2 \rightleftharpoons 2NO$	−14096.	−0.6893	−1375.3	9.668
$2N \rightleftharpoons N_2$	108142.	−1.744	−3558.2	0.595

Tables E.1, E.2, and E.3 for c_p, h, and ϕ in Appendix E give tabulated values of these functions at 100 K intervals in the range between 300 and 1000 K and at 500 K intervals between 1000 and 6000 K.

A more ambitious scheme for computing thermodynamic properties of gases will be found in Prothero, "Computing with Thermochemical Data" *Combustion and Flame*, Vol. 13, No. 4, 1969. The expressions are derived starting with a seventh-order polynomial for c_p. Fourteen coefficients are listed for 47 gases: seven in the range 300 to 2000 K, and seven in the range 2000 to 6000 K. The c_p values were drawn from the JANAF tables.* With c_p calculable over the range 300 to 6000 K, values of enthalpy and entropy are obtained by integration. Prothero lists values of enthalpy and entropy at 300 K for each gas. These are the integration constants in Eqs. C-1 and are chosen to incorporate the heats of formation. With a datum value for the reaction constant $K_p(300)$, reaction constants can then be found.

The system developed by Prothero is more accurate than the equations listed in C-3. For classroom purposes, however, the latter system is entirely satisfactory.

APPENDIX D   FURTHER DISCUSSION OF CHEMICAL EQUILIBRIUM

In Section 10.8, LeChatelier's equation, which reveals the direction in which the equilibrium composition shifts when changes in pressure and temperature are imposed on a system, the quantity Ψ appears, defined in Eq. 10-26 (repeated here)

$$\frac{1}{\Psi} = \Sigma \frac{\nu_i^2}{x_i} - (\Delta\nu_i)^2 \qquad \text{D-1}$$

In Section 10.8, it was shown that stability requires that Ψ be positive, a result arrived at by physical reasoning. A mathematical proof is given here. Also, chemical equilibrium will be examined in a broader context than was presented in Sections 10.3 and 10.4.

We consider the behavior of a system that is at P and T, located in surroundings (e.g., the atmosphere) that are at P_0 and T_0. The system contains four gases that engage in a reversible reaction

$$\nu_1 A_1 + \nu_2 A_2 \rightleftharpoons \nu_3 A_3 + \nu_4 A_4 \qquad \text{D-2}$$

and a fifth gas that is inert. The system is closed to mass transfer, but heat and work transfers may take place between system and surroundings.

The system and surroundings constitute a universe, in the thermodynamic sense, and the second law requires that

$$dS + dS_o \geqslant 0 \qquad \text{D-3}$$

where dS and dS_o are the system and surroundings entropy changes, respectively. The equality holds at equilibrium; otherwise the inequality applies. Equation D-3 is the

---

Joint Army Navy Air Force (JANAF) Thermochemical Tables, prepared by Dow Chemical Co. Distributed by the Clearinghouse for Federal Scientific and Technical Information, Publication PB 168370, 1970.

starting point for all thermodynamic discussions of equilibrium. Define heat transfer in the direction from surroundings to system as positive, and suppose that heat transfer of amount δQ occurs. Because the surroundings are infinite in extent, dS_o can be replaced by

$$dS_o = -\frac{\delta Q}{T_o}$$

and δQ may be replaced by

$$\delta Q = dU + P\,dV$$

so that Eq. D-3 becomes

$$dU + P\,dV - T_o\,dS \leqslant 0 \qquad\qquad \text{D-4}$$

We now consider three cases of constraints on the system. (Other combinations of constraints can be concocted, even bizarre ones, but they cannot be realized in a practical sense and will not be discussed.) The three are

> I Constant U and V
> II Constant P and T
> III Constant V and T

The first is an isolated system; the second, a piston and uninsulated cylinder; and the third, a rigid, uninsulated container. These are the only practical possibilities.

Case I: Constant U and V, an isolated system. This has already been discussed in Section 10.3, where it was shown that at equilibrium the equation

$$\nu_1(h_1 - Ts_1) + \nu_2(h_2 - Ts_2) = \nu_3(h_3 - Ts_3) + \nu_4(h_4 - Ts_4)$$

must be satisfied.

Case II: Constant P and T. Note first that we must have $P = P_o$ and $T = T_o$. If these conditions are not satisfied, then in time they will be reached, by way of heat and work transfers. Equation D-4 now becomes

$$d(H - TS) \leqslant 0 \qquad\qquad \text{D-5}$$

The combination $H - TS$ is known as the *Gibbs function*, G, or the *free energy*, and Eq. D-5 states that G will decrease as the system moves toward equilibrium and will be a minimum at equilibrium.

Now, for ideal gases,

$$H = \Sigma n_i h_i$$
$$S = \Sigma n_i s_i$$
$$h_i = \int_0^T c_{p_i}\,dT + h(0)$$

and

$$s_i = \int_0^T \frac{c_{pi}}{T} \, dT - R \ln(p_i) + s_i(0)$$

or

$$s_i = \int_0^T \frac{c_{pi}}{T} dT - R \ln(P) + s_i(0) - R \ln(x_i)$$

so that the Gibbs function can be written in the form

$$G = \Sigma n_i \, \mu_i, \qquad \mu_i = h_i - T s_i$$

where the μ_i are referred to as the *chemical potentials* and have the form

$$\mu_i = \mu_i^{\,o}(P, T) + RT \ln(x_i) \qquad\qquad \text{D-6}$$

which separates μ_i in a function of P and T, and another in x_i, the mole fraction. Note that $\mu_i^{\,o}$ is the chemical potential of the pure gas, $x_i = 1$. Finally, introduce the *extent of reaction*, ϵ, to describe the mole numbers of the four reactive gases, as in Eq. 10-31,

$$n_1 = \nu_1(1 - \epsilon) + a_1, \qquad n_3 = \nu_3 \epsilon + a_3$$
$$n_1 = \nu_2(1 - \epsilon), \qquad\quad n_4 = \nu_4 \epsilon \qquad\qquad \text{D-7}$$

where a_2 and a_4 have been set equal to zero. (Of course, a_1 or a_3 or both could also be zero.)

We can now determine the conditions for which G is stationary and a minimum. The condition for stationary G is

$$\left(\frac{\partial G}{\partial \epsilon}\right)_{P, T} = 0$$

or

$$\left(\frac{\partial G}{\partial \epsilon}\right)_{P, T} = \Sigma\left[n_i\left(\frac{\partial \mu_i}{\partial \epsilon}\right)_{P, T} + \mu_i \frac{dn_i}{d\epsilon}\right] = 0 \qquad\qquad \text{D-8}$$

The first term is zero because

$$\Sigma n_i\left(\frac{\partial \mu_i}{\partial \epsilon}\right)_{P, T} = \Sigma n_i \frac{\partial}{\partial \epsilon}\left[\mu_i^{\,o}(P, T) + RT \ln(x_i)\right]_{P, T}$$

$$= RT \, \Sigma n_i \frac{d \ln(x_i)}{d\epsilon}$$

$$= RT \, \Sigma \frac{n_i}{x_i} \frac{dx_i}{d\epsilon}$$

$$= NRT \frac{d}{d\epsilon} \Sigma(x_i) = 0$$

Equation D-8 reduces to

$$\Sigma \mu_i \frac{dn_i}{d\epsilon} = 0$$

or

$$\nu_1\mu_1 + \nu_2\mu_2 = \nu_3\mu_3 + \nu_4\mu_4 \qquad\qquad\qquad \text{D-9}$$

and since $\mu_i = h_i - Ts_i$, the criterion reduces to Eq. 10-10, as in the case for stationary entropy. The two approaches to the discussion of chemical equilibrium, using different constraints, reach the same conclusion. Equilibrium is an internal condition, a state of the system. The constraints on the system play a role in fixing the pressure, temperature, and composition of the system. Equation D-9 must be satisfied, regardless of the choice constraints.

Differentiate Eq. D-8 a second time, and obtain

$$\left(\frac{\partial^2 G}{\partial\epsilon^2}\right)_{P,T} = \Sigma n_i \left(\frac{\partial^2 \mu_i}{\partial\epsilon^2}\right)_{P,T} + 2\Sigma \frac{dn_i}{d\epsilon}\left(\frac{\partial\mu_i}{\partial\epsilon}\right)_{P,T} + \Sigma\mu_i \frac{d^2 n_i}{d\epsilon^2}$$

The last term is zero because the mole numbers are linear in ϵ. Carry out the differentiation indicated in the first two terms, and arrive at

$$\left(\frac{\partial^2 G}{\partial\epsilon^2}\right)_{P,T} = \frac{RT}{N\Psi} \qquad\qquad\qquad \text{D-10}$$

where Ψ is defined as in Eq. D-1. The second derivative at the minimum point of a function must be positive. Since N, R, and T are positive, Ψ is positive.

As an illustration of the G versus ϵ curves, consider the dissociation of oxygen

$$O_2 \rightleftharpoons 2O$$

The mole numbers are

$$n_{O_2} = 1 - \epsilon, \qquad n_O = 2\epsilon, \qquad N = 1 + \epsilon$$

so that

$$G = \left[(1 - \epsilon)u_{O_2}^o + 2\epsilon\mu_O^o\right] + RT\left[(1 - \epsilon)\ln\frac{1 - \epsilon}{1 + \epsilon} + 2\ln\frac{2\epsilon}{1 + \epsilon}\right]$$

Figure D.1 shows the second term in square brackets plotted against ϵ, along with Ψ on the same plot. The first term in square brackets is linear in ϵ, and its slope depends on the relative values of $\mu_{O_2}^o$ and μ_O^o, which are functions of total pressure and temperature. Figure D.2 shows how the equilibrium point shifts with $\mu_{O_2}^o$ and μ_O^o.

Case III: Constant V and T. The pressure of the surroundings can take on any value, but the temperature of the system and the surroundings must be the same.

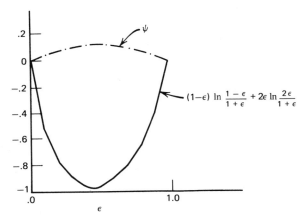

Figure D.1

Setting $T_o = T$ in Eq. D-4,

$$d(U - TS) \leqslant 0 \qquad \text{D-11}$$

The combination $U - TS$ is the *Helmholtz function* (it is also referred to as the *free energy*, which is confusing), denoted by F. The treatment follows the same pattern as in the case of the Gibbs function. Set

$$F = \Sigma n_i f_i \qquad \text{D-12}$$

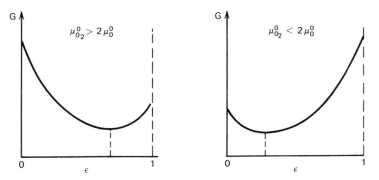

Figure D.2 Gibbs function, G, versus degree of reaction, ϵ, for a mixture of O_2 and O. Pressure and temperature fix the chemical potentials, $\mu^o_{O_2}$ and μ^o_O. The reaction is $O_2 \rightleftharpoons 2O$.

and derive an expression for f_i consisting of a function of V and T and whatever is left over. For ideal gases,

$$u_i = \int_0^T c_{v_i}\, dT + u_i(0)$$

$$s_i = \int_0^T c_{p_i}\, \frac{dT}{T} - R \ln \frac{RT}{V} + s_i(0) - R \ln(n_i)$$

so that

$$f_i = f_i^o(V, T) + RT \ln(n_i)$$

and

$$f_i^o = u_i - T\phi_i + RT \ln \frac{RT}{V} - Ts_i(0)$$

Setting

$$\left(\frac{\partial F}{\partial \epsilon}\right)_{V,T} = 0$$

the result will reduce to Eq. D-9. A second differentiation leads to

$$\left(\frac{\partial^2 F}{\partial \epsilon^2}\right)_{V,T} = \frac{RT}{N}\left[\frac{\nu_1^2}{x_1} + \frac{\nu_2^2}{x_2} + \frac{\nu_3^2}{x_3} + \frac{\nu_4^2}{x_4}\right]$$

which is positive.

Note that in carrying out the differentiations in Cases II and III, the summations extend over the four reactive gases and the inert gas. Note also that, with n_x denoting the moles of inert gas,

$$\frac{dn_x}{d\epsilon} = 0, \qquad \text{but} \quad \frac{dx_x}{d\epsilon} \neq 0$$

since the mole fraction of the inert gas, x_x, involves the total mole number, which is not constant, for the general case.

The four functions

$$U, \qquad\qquad dU = T\, dS - P\, dV$$
$$H = U + PV, \qquad dH = T\, dS + V\, dP$$
$$G = H - TS, \qquad dG = V\, dP - S\, dT$$
$$F = F - TS, \qquad dF = -P\, dV - S\, dT$$

are the only *useful* thermodynamic functions that can be derived from linear combinations of U, PV, and TS. For example,

$$X = U - PV$$

is a thermodynamic function, which simply means that values of X can be assigned in terms of T and P, or T and V, or P and V. This is another way of stating that the value of X depends on the state of the system, not on how that state was reached from some other state. But

$$dX = dU - d(PV) = T\,dS - P\,dV - P\,dV - V\,dP$$
$$= T\,dS - 2P\,dV - V\,dP$$

The function $U - PV$ does not have a differential that can be written simply in terms of a pair selected from the group (dS, dT, dP, dV). In the above equation, dS, dV, and dP are not independent; one must be eliminated in terms of the other two. The expression for dX is therefore not simple, as is the case for dU, dH, dG, and dF. Consequently, U, H, F, and G are useful, but X (and others that can be formulated) is not.

APPENDIX E TABLES OF THERMODYNAMIC PROPERTIES

The tables in this appendix were adapted from NACA TN 1037 (1951) "General Method and Thermodynamic Tables for Computation of Equilibrium Composition and Temperature of Chemical Reactions." (NACA stands for National Advisory Committee on Aeronautics, the forerunner of NASA.)

Table E.1 c_p^o, constant pressure heat capacity at one atmosphere (kJ/kmol K)
Table E.2 h^o, enthalpy at one atmosphere (kJ/kmol)
Table E.3 ϕ^o, entropy at one atmosphere (kJ/kmol K)
Table E.4   Some heats of reaction, H_{rp}, and standard heats of formation, Δh_f^o
Table E.5 $\text{Log}_{10} K_p$, the reaction constant, for several formation reactions.

Since c_p (and c_v) is a weak function of pressure, the one-atmosphere values may be used at all pressures without introducing appreciable error.

Note: As pointed out in Chapter 2, the enthalpy tables incorporate the appropriate heats of formation.

In constructing enthalpy tables for gases, the heat of formation occupies a role similar to that of the heat of vaporization in the tabulation of enthalpy values in, say, the steam tables. For example, suppose that enthalpy tables of liquid H_2O and vapor H_2O had been prepared, each with datum values chosen without regard for one another, and further suppose that these tables are to be used to evaluate

$$h(p_2, t_2) - h(p_1, t_1)$$

where point 1 is in the liquid region, and point 2 is in the vapor region. Because of the arbitrary datum selections, the enthalpy difference computed from tabulated values will give a totally meaningless result. To eliminate ambiguity, we have to proceed by writing the difference in an expanded form, to wit

$$h(p_2, t_2) - h_g(t_o) + h_g(t_o) - h_f(t_o) + h_f(t_o) - h(p_1, t_1)$$

with subscripts f and g denoting, respectively, saturated liquid and saturated vapor.

Regrouping the terms

$$[h(p_2, t_2) - h_g(t_o)] + h_{fg}(t_o) - [h(p_1, t_1) - h_f(t_o)]$$

where

$$h_{fg}(t_o) = h_g(t_o) - h_f(t_o)$$

$$= \text{heat of vaporization at } t_o.$$

The pairs of terms within square brackets are unambiguous, which is to say they are independent of the datum values in the two tables. Also, $h_{fg}(t_o)$ is a *measured* quantity or is derived from measurements.

This is a cumbersome and tedious routine. It can be eliminated, and is eliminated in steam tables, by choosing a single arbitrary datum value for the enthalpy of H_2O in some state (usually chosen to avoid negative values) and then connecting the liquid and vapor regions so that values of $h_f(t)$ and $h_g(t)$ comply with

$$h_{fg}(t) = h_g(t) - h_f(t)$$

When steam tables are so arranged, *all* enthalpy differences can be evaluated by simple subtraction.

In constructing enthalpy tables for gases, the heat of formation is employed to connect tables that involve the same atoms or groups of atoms. If, for example, tabulated values for $h_{O_2}(t)$ have been computed, with an arbitrary datum, then a table for atomic oxygen $h_O(t)$ can be secured by setting

$$h_O(298 \text{ K}) = \tfrac{1}{2}(\Delta h_f^o - h_{O_2})_{298 \text{ K}}$$

relating all $h_O(t)$ values to $h_O(298 \text{ K})$. The advantage that accompanies this special choice for the atomic oxygen table is that an enthalpy difference, such as

$$2h_O(t_1) - h_{O_2}(t_2)$$

can be evaluated with values from the two tables and has unambiguous physical meaning. Note, of course, that

$$h_O(t_1) - h_{O_2}(t_2)$$

has no physical meaning, since the two terms represent different masses of oxygen.

In this way the enthalpy of C, H_2, O_2, and N_2 can be assigned independently, since no element can be formed from any other or any combination of others. Now enthalpy assignments can be made to all atoms and molecules that contain C, H, O, and N atoms, provided that the appropriate heat of formation has been measured or calculated with the methods of quantum mechanics. (The only precaution regularly observed is to make the arbitrary enthalpy assignments so that no atom or molecule appears with negative enthalpy, which, while legitimate and entirely meaningful, merely invites computational errors.)

The advantage is that enthalpy differences of the form

$$ah_A(t_A) + bh_B(t_B) + \ldots - xh_X(t_X) - yh_Y(t_Y) - \ldots$$

are meaningful, provided

$$aA + bB + \ldots$$

and

$$xX + yY + \ldots$$

contain similar numbers of similar atoms, which is to say that one set of mole numbers and chemical species can be formed by one or more reactions from the other set. Enthalpy differences must observe conservation of matter.

Everything that has been said here with reference to enthalpy applies also to internal energy and entropy. (In the case of entropy, the third law establishes the notion of absolute entropy value.)

TABLE E.1
c_p^o (kJ/kmol K) FOR GASES

T (K)	CO	CO_2	H	H_2	H_2O	N
300	29.14	37.21	20.79	28.85	33.58	20.79
400	29.34	41.30	20.79	29.18	34.25	20.79
500	29.79	44.61	20.79	29.26	35.21	20.79
600	30.44	47.33	20.79	29.32	36.30	20.79
700	31.17	49.58	20.79	29.43	37.48	20.79
800	31.90	51.46	20.79	29.61	38.72	20.79
900	32.58	53.04	20.79	29.87	39.99	20.79
1000	33.19	54.37	20.79	30.20	41.26	20.79
1100	33.71	55.50	20.79	30.58	42.45	20.79
1200	34.17	56.44	20.79	30.99	43.57	20.79
1300	34.58	57.24	20.79	31.42	44.63	20.79
1400	34.93	57.92	20.79	31.86	45.64	20.79
1500	35.23	58.53	20.79	32.30	46.58	20.79
2000	36.25	60.68	20.79	34.29	50.24	20.79
2500	36.84	61.99	20.79	35.83	52.53	20.82
3000	37.23	62.94	20.79	37.06	54.03	20.97
3500	37.50	63.66	20.79	38.13	55.01	21.28
4000	37.72	64.25	20.79	39.09	55.68	21.82
4500	37.90	64.75	20.79	39.97	56.17	22.56
5000	38.06	65.21	20.79	40.79	56.57	23.48
5500	38.20	65.67	20.79	41.55	56.96	24.50
6000	38.33	66.13	20.79	42.26	57.34	25.60

T(K)	NO	N_2	O	OH	O_2
300	29.85	29.12	21.90	29.87	29.38
400	29.97	29.25	21.48	29.60	30.11
500	30.50	29.58	21.26	29.49	31.09
600	31.25	30.11	21.12	29.51	32.09
700	32.04	30.76	21.04	29.65	32.98
800	32.77	31.43	20.98	29.92	33.74
900	33.43	32.10	20.94	30.27	34.36
1000	34.00	32.70	20.91	30.68	34.88
1100	34.49	33.25	20.89	31.13	35.31
1200	34.90	33.74	20.88	31.59	35.68
1300	35.25	34.16	20.86	32.06	36.00
1400	35.56	34.53	20.85	32.52	36.29
1500	35.82	34.85	20.84	32.95	36.56
2000	36.70	35.99	20.83	34.76	37.78
2500	37.22	36.65	20.85	36.04	38.92
3000	37.58	37.07	20.94	36.98	39.96
3500	37.86	37.38	21.09	37.72	40.84
4000	38.10	37.61	21.30	38.33	41.56
4500	38.32	37.80	21.55	38.87	42.09
5000	38.53	37.97	21.80	39.35	42.49
5500	38.73	38.13	22.05	39.79	42.79
6000	38.93	38.28	22.29	40.18	43.01

TABLE E.1

EXPRESSIONS FOR THE SPECIFIC HEAT OF GASES AND VAPORS THAT ARE ADEQUATE FOR USE IN INTEGRALS OF THE TYPE OCCURRING IN EQS. 2-41 AND 2-42.

$c_p = a + bT$ (kJ/kmol K) $298 < T < 900$ K

Gas	a	b	Molecular Weight
CO	27.4	0.0058	28
CO_2	28.8	0.028	44
H_2	28.3	0.0019	2
H_2O	30.5	0.0103	18
N_2	27.6	0.0051	28
O_2	27.0	0.0079	32
Air	27.5	0.0057	28.97
CH_4	20.1	0.0052	16
C_2H_6	52.3	0.104	30
C_3H_8	25.3	0.162	44
C_8H_{18}	38.4	0.429	114
$C_{10}H_{22}$	38.1	0.656	142
CH_3OH	8.37	0.126	32
C_2H_5OH	18.8	0.159	46
C_6H_6	27.2	0.218	78

TABLE E.2
h^o (kJ/kmol) FOR GASES

T(K)	CO	CO_2	H	H_2	H_2O
298.16	283 754	9 364	363 212	290 540	57 316
300	283 807	9 433	363 250	290 593	57 376
400	286 729	13 367	365 329	293 499	60 773
500	289 684	17 688	367 408	296 421	64 252
600	292 694	22 269	369 487	299 346	67 836
700	295 775	27 118	371 566	302 287	71 557
800	298 930	32 172	373 645	305 241	75 399
900	302 155	37 405	375 724	308 214	79 353
1000	305 443	42 769	377 803	311 217	83 425
1100	308 788	48 262	379 879	314 257	87 610
1200	312 182	53 860	381 958	317 335	91 911
1300	315 620	59 543	384 036	320 456	96 321
1400	319 096	65 301	386 115	323 620	100 835
1500	322 604	71 124	388 194	326 828	105 446
2000	340 499	100 974	398 587	343 493	129 722
2500	367 154	131 662	408 980	361 039	155 455
3000	377 309	162 907	419 373	379 373	182 117
3500	396 000	194 565	429 766	398 076	209 392
4000	414 806	226 547	440 159	417 383	237 073
4500	433 713	258 798	450 552	437 150	265 039
5000	452 704	291 288	460 945	457 340	293 225
5500	471 769	324 009	471 338	477 926	321 609
6000	490 903	356 919	481 731	498 879	350 186

TABLE E.2 (CONTINUED)
h^o (kJ/kmol) FOR GASES

$T(K)$	N	N_2	NO	O	OH	O_2
298.16	365 894	15 780	106 855	256 109	195 948	17 200
300	365 932	15 834	106 907	256 149	190 003	17 254
400	368 011	18 751	109 895	257 061	198 975	20 225
500	370 090	21 691	112 084	260 452	201 926	23 284
600	371 974	24 674	116 005	262 570	204 880	26 444
700	374 247	27 717	120 421	264 678	207 834	29 699
800	376 326	30 827	122 400	266 779	210 811	33 036
900	378 404	34 004	125 709	268 876	213 819	36 442
1000	380 483	37 245	129 079	270 968	216 868	39 904
1100	382 562	40 543	132 504	273 058	219 958	43 414
1200	384 640	43 892	135 973	275 147	223 094	46 963
1300	386 719	47 287	139 481	277 234	226 277	50 547
1400	388 797	50 721	143 022	279 320	229 506	54 161
1500	390 876	54 190	146 590	281 405	232 779	57 804
2000	401 270	71 929	164 740	291 821	249 731	76 394
2500	411 671	90 101	183 229	302 238	267 449	95 569
3000	422 114	108 539	201 931	312 682	285 713	115 294
3500	432 667	127 156	220 793	323 186	304 394	135 501
4000	443 433	145 906	239 785	333 783	323 411	156 108
4500	454 520	164 760	258 891	344 494	342 715	177 025
5000	466 024	183 705	278 102	355 331	362 274	198 176
5500	478 015	202 732	297 415	366 293	382 060	219 501
6000	490 537	221 837	316 829	377 379	402 054	240 954

Also, h^o (sat. liquid H_2, 20 K) = 282 612 (kJ/kmol)
h^o (sat. liquid O_2, 90 K) =   4 315 (kJ/kmol)
h^o (solid C, 298.16 K)   = 385 346 (kJ/kmol)

TABLE E.3

$$\phi^o = \int_0^T c_p^o \frac{dT}{T} = \text{ABSOLUTE ENTROPY AT 1 atm (kJ/kmol K) FOR GASES}$$

$T(K)$	CO	CO_2	H	H_2	H_2O	N
298.16	197.91	213.64	114.611	130.59	188.72	153.195
500	213.08	234.78	125.357	145.64	206.46	163.941
1000	234.789	269.161	139.764	166.122	232.706	178.348
1500	248.677	292.098	148.192	178.752	250.491	186.776
2000	258.967	309.258	154.172	188.327	264.433	192.756
2500	267.125	322.948	158.810	196.151	275.907	197.399
3000	273.879	334.339	162.600	202.797	285.625	201.206
3500	279.639	344.098	165.804	208.591	294.032	204.459
4000	284.661	352.638	168.580	213.746	301.424	207.334
4500	289.115	360.235	171.028	218.401	308.011	209.944
5000	293.116	367.081	173.218	222.655	313.950	212.368
5500	296.751	373.318	175.199	226.579	319.361	214.653
6000	300.080	379.052	177.008	230.225	324.333	216.831

$T(K)$	N_2	NO	O	O_2	OH
298.16	191.49	210.62	160.954	205.06	183.63
500	206.63	226.14	172.091	220.61	198.96
1000	228.066	248.404	186.683	243.505	219.622
1500	241.778	262.579	195.147	257.997	232.494
2000	251.976	273.016	201.140	268.685	242.235
2500	260.083	281.265	205.789	277.238	250.137
3000	266.805	288.084	209.597	284.427	256.794
3500	272.545	293.898	212.835	290.655	262.552
4000	277.551	298.970	215.665	296.158	267.630
4500	281.993	303.471	218.188	301.084	272.177
5000	285.985	307.519	220.471	305.541	276.298
5500	289.611	311.200	222.561	309.606	280.069
6000	292.936	314.578	224.490	313.339	283.548

TABLE E.4
H_{rp}, CONSTANT PRESSURE HEAT OF REACTION, kJ/kmol
PRODUCTS ARE GASEOUS (298.16 K)

		Molecular Weight	Liquid Fuel	Gaseous Fuel
Methane	CH_4	16		−797 570
Ethane	C_2H_6	30		−1 419 900
Propane	C_3H_8	44	−2 016 900	−2 032 800
n-Butane	C_4H_{10}	58	−2 622 800	−2 644 200
n-Pentane	C_5H_{12}	72	−3 228 200	−3 254 700
n-Hexane	C_6H_{14}	86	−3 834 600	−3 866 000
n-Heptane	C_7H_{16}	100	−4 441 100	−4 477 500
n-Octane	C_8H_{18}	114	−5 047 800	−5 089 100
n-Nonane	C_9H_{20}	128	−5 654 700	−5 700 700
n-Decane	$C_{10}H_{22}$	142	−6 261 300	−6 312 300
Methyl alcohol	CH_3OH	32	−814 460	−851 840
Ethyl alcohol	C_2H_5OH	46	−1 498 800	−1 541 000
Benzene	C_6H_6	78	−3 399 500	−3 433 400

For H_2O: $h_{fg}(25°C, 298.16 \text{ K}) = 43\ 956$ kJ/kmol.

TABLE E.4 (CONTINUED)
Δh_f^o, STANDARD ENTHALPY OF FORMATION, kJ/mol
(298.16 K)

	Molecular Weight	Δh_f^o
Carbon monoxide, CO	28	−110 000
Carbon dioxide, CO_2	44	−393 000
Water vapor, H_2O	18	−241 800
Nitroglycerin, $C_3H_8(NO_3)_3$	227	−346 000
Urea, $(NH_2)_2CO$	60	−323 000
TNT, $C_6H_2CH_3(NO_2)_3$	227	−54 400
Cellulose, $(C_6H_{10}O_5)_n$	162n	−950 000n
Explosive D, $C_6H_2(NO_2)_3O\ NH_4$	246	−343 000
Glycerin, $C_3H_5(OH)_3$	92	−657 000
Ethyl urea, $CO\ NH_2 \cdot NH \cdot C_2H_5$	88	−340 000
Ethyl acetate, $C_2H_5 \cdot CH_3 \cdot COO$	88	−480 700
Cellulose decanitrate, $C_{24}H_{30}O_{20}(NO_2)_{10}$	1098	−2 803 000

TABLE E.5
LOG$_{10}$ K_p FOR SEVERAL FORMATION REACTIONS (PRESSURES IN ATMOSPHERES)

T (K)	C + O ⇌ CO	C + 2O ⇌ CO$_2$	2H ⇌ H$_2$	2H + O ⇌ H$_2$O	H + O ⇌ OH	2O ⇌ O$_2$	2N ⇌ N$_2$	O + N ⇌ NO
298.16	182.2536	267.6053	71.2098	151.5648	69.3677	80.6182	119.4348	84.8403
1000	49.5767	69.4953	17.2883	37.0674	16.9801	19.4400	31.0841	21.1937
1500	30.6703	41.3016	9.5105	20.5727	9.4213	10.6752	18.4526	12.0721
2000	21.1878	27.1855	5.5798	12.2533	5.6027	6.2695	12.1063	7.4848
2500	15.4834	18.7104	3.2018	7.2289	3.2931	3.6157	8.2835	4.7193
3000	11.6713	13.0585	1.6064	3.8617	1.7434	1.8415	5.7261	2.8677
3500	8.9417	9.0207	0.4610	1.4458	0.6306	0.5718	3.8933	1.5397
4000	6.8892	5.9919	−0.4012	−0.3732	−0.2080	−0.3818	2.5138	0.5395
4500	5.2885	3.6356	−1.0736	−1.7933	−0.8630	−1.1243	1.4365	−0.2422
5000	4.0043	1.7500	−1.6126	−2.9335	−1.3890	−1.7188	0.5703	−0.8708
5500	2.9505	0.2067	−2.0539	−3.8695	−1.8209	−2.2058	−0.1426	−1.3884
6000	2.0696	−1.0800	−2.4216	−4.6522	−2.1820	−2.6121	−0.7410	−1.8228

SUBJECT INDEX

DEFINITIONS OF VARIOUS SYMBOLS

Definitions of Various Symbols

abo, rocket acceleration at burnout, 194

AC, AH, AN, AO, abundance of carbon, hydrogen, nitrogen and oxygen atoms, respectively, 272

aig, retrorocket acceleration at ignition, 203

ao, rocket acceleration at liftoff, 194

b, co-volume factor, 319

c, molecular velocity, 250, 325

C_p^*, constant pressure heat capacity of a gas mixture maintained in chemical equilibrium, 299, 334

C_p, C_v, heat capacities (kJ/K), 334

c_p, c_v, heat capacities (kJ/kg K or kJ/kmol K), 334

CR, compression ratio, 64, 149, 176

F/A, fuel/air ratio (kg/kg or kmol/kmol, both are used)

h, enthalpy (kJ/kmol or kJ/kg), 332

H, enthalpy (kJ), 334

H_{rp}, constant pressure heat of reaction, 2

I_{sp}, specific impulse, 199

$k = C_p/C_v = c_p/c_v$, heat capacity ratio, 332

K_p, reaction constant, 255

$\dot{M}$, mass flow rate in rockets, 197

MEP, mean effective pressure, 83

$N(I)$, mole numbers (FORTRAN subscript notation), 10

n, ratio of burned gas/total gas in progressive burning, 98

p_i, partial pressure, 334

p, gas pressure exerted on a moving piston, 325

P, total (statical) pressure, 334

PR, pressure ratio function in gas turbine analysis, 39
Q_p, heat release, constant pressure combustion, 13
Q_v, heat release, constant volume combustion, 16
R, gas constant, 332
s, entropy (kJ/kmol K or kJ/kg K), 333
S, entropy, (kJ/K), 334
S, normal burning velocity, 113
S_f, flame speed, 113
SR, volume relation in two stroke engines, 178
u, internal energy (kJ/kmol or kJ/kg), 332
U, internal energy (kJ), 334
U_{rp}, constant volume heat of reaction, 4
v_s, surface regression velocity in powders, 322
Va, acoustic velocity, 198, 314
Vbo, rocket velocity at burnout, 191, 210
Vco, velocity for circular orbit, 215
Vig, retrorocket velocity at ignition, 202
VR, volume relation in two stroke engines, 178
Y, oxygen/fuel mole ratio, 10
YCC, chemically correct oxygen/fuel mole ratio in C/H/N/O system, 10
$YMIN$, minimum oxygen/fuel mole ratio for complete combustion C/H/N/O systems, 10
zbo, rocket altitude at burnout, 191
zig, retrorocket altitude at ignition, 202
$zmax$, maximum rocket altitude, 191, 192
β, propellant burning rate in chemical rockets, 190
Δ, powder loading factor in guns, 326
Δh_f°, standard heat of formation, 6
ΔH, heat of reaction for a reversible reaction, 256
$\Delta\theta_c$, duration of combustion in Otto engines, 110
ϵ, extent of reaction, 264
ϵ, regenerator effectiveness, 44
η_c, compressor efficiency, 41
η_t, turbine efficiency, 41
η_{th}, thermal efficiency, 41
  $= w/U_{rp}$ for constant volume combustion engines
  $= w/H_{rp}$ for constant pressure combustion engines
  w = net work per unit of fuel burned
  (But *see* 179, two thermal efficiencies can be defined for two stroke engines).
η_v, volumetric efficiency, 129
θ, crank angle, measured from bottom dead center, for reciprocating engines, 65
ν, stoichiometric coefficient, 252
$\Delta\nu$, stoichiometric coefficient change for a reversible reaction, 256
π, propellant loading in chemical rockets, 190
ϕ, temperature part of the entropy function, 333
Ψ, mobility of a gas mixture in chemical equilibrium, 263